Visual Revelations

Visual Revelations

Graphical Tales of Fate and Deception from Napoleon Bonaparte to Ross Perot

Howard Wainer

COPERNICUS
AN IMPRINT OF SPRINGER-VERLAG

Published in the United States by Copernicus,
an imprint of Springer-Verlag New York, Inc.

Copernicus
Springer-Verlag New York, Inc.
175 Fifth Avenue
New York, NY 10010

About the cover: "Chart of the National Debt of England," which appeared as
plate 20, opposite p. 83 in William Playfair's *Commercial and Political Atlas*
(third edition), published in London in 1801. This beautifully executed line chart
uses the innovation of an irregularly spaced grid along the time (horizontal) axis
to demark events of important economic consequence. The inexorable conclusion
we draw is that war was bad for England's national debt.

Library of Congress Cataloging-in Publication Data

Wainer, Howard.
    Visual revelations: graphical tales of fate and deception from
    Napoleon Bonaparte to Ross Perot / by Howard Wainer.
        p.   cm.
    Includes bibliographical references (p.   –   ) and index.
    ISBN 0-387-94902-X (hardcover : alk. paper)
    1. Mathematical statistics—Graphic methods.   I. Title.
    QA276.3.W35   1997
    001.4'266—dc20   96-44235

Manufactured in the United States of America.
Printed on acid-free paper.
9 8 7 6 5 4 3 2 1

ISBN 0-387-94902-X   SPIN 10557017

To Linda and Sam

Contents

———✦———

Acknowledgments

⟺◆⟹

The writing of this book spans almost twenty-five years. In that time an enormous number of intellectual debts have accumulated. I am delighted to be able to acknowledge my gratitude to the many who have contributed to my interest in and understanding of graphics.

The path began with John W. Tukey, who asked me to act as a discussant at a symposium on exploratory data analysis he had organized in 1972. Feeling that discussing the statistical technology of the foremost leaders in that field was an overreach for someone only barely out of graduate school, I decided to center my discussion on the need to verify the efficacy of graphical innovations experimentally. Fred Mosteller, who was also part of the symposium, suggested that I publish my findings. I did, and thus my graphical career began. That article came to the attention of Albert Biderman, who was looking for someone to run a research project at the Bureau of Social Science Research on the uses of graphics for social reporting. For many reasons this seemed like an interesting path to follow, so I agreed enthusiastically to join, and moved to Washington. One product of this project was the translation of Jacques Bertin's monumental *Semiologie Graphique* into English, on which Bill Berg and I collaborated happily. The ideas I absorbed from Bertin and Tukey remain two of the principal legs supporting the seat of my graphical experience.

The third leg is Edward Tufte. I first met Edward when he gave a lecture on improving data display at the University of Chicago in 1977. Although his thinking on graphics was still at an early stage, his ideas had a profound influence on me; figures 8, 9, and 10 in chapter 1 all came from that 1977 lecture. Many of the ideas and examples in this book (especially chapter 1) have their origins with Tufte. I am grateful indeed for his permission to reprint his figures. There is no one whose work I admire more than Tufte's. His three books (1983, 1990, 1997) are a wonderful and beautiful manifestation of his thoughts and taste.

During the course of the Graphics Social Reporting Project (1977–1980), I met with many of the principal workers in statistical graphics, but my friend and colleague Albert Biderman was, and remains, enormously influential. Indeed, some of the prose in the introduction to section I as well as the introduction to chapter 5 had its origin in a piece that we cowrote; I am pleased that Al has given his permission to reuse it here. Many of the ideas presented in this book are my interpretations of his Delphic suggestions. Of course, he should not be held responsible for how I've used them.

Since 1980 I have been employed by the Educational Testing Service. Although only a little of my work there is directly related to data display, I have always been accorded time and support to pursue my graphical interests. Without the forbearance of my employer this book would never have been completed, and to that I am indebted. I would specifically like to express my appreciation to the trustees of the Educational Testing Service, who through the 1990 Senior Scientist Award afforded me additional freedom to pursue various graphical notions. In the preface to his *Sketches from a Life*, the storied diplomat and historian George Kennan quoted Anton Chekov, a doctor, who wrote that while medicine was his wife, literature was his mistress. Kennan added that his situation was similar, but his mistress was neither as beautiful nor as varied as Chekov's. My relationship with statistics and graphics follows in the same pattern, although trailing far behind in breadth and depth as well as beauty.

In many books one often sees statements like "To Zelda, my long-suffering wife, for late dinners and her understanding that following my muse precludes doing the dishes." Well, I would like to thank my wife, Linda Steinberg, for her intolerance. She has read everything in here, as well as much that happily is not. She has been intolerant of pompous or unclear prose, of sloppy thinking, and of pointless discussion. Her contribution, in short, was not to the infrastructure that allowed me to write this book, but rather to its intellectual content. My gratitude, which is immense, is for her direct contributions toward making this a better book. Any late dinners I usually ended up cooking myself.

Over long periods of time, a number of my colleagues have provided bits and pieces of help, which when aggregated have become substantial. Principal among these is my former student, long-term friend, and frequent collaborator, David Thissen. We have written so much together that it is often impossible to determine on whose word processor the prose, and indeed the ideas, originated. I believe that the prose in this book is mostly mine, but some of the ideas are surely David's. Happily, he has so many ideas that he doesn't begrudge my borrowing a few.

In addition, I would like to thank for their comments, suggestions, and contributions to various pieces Jill Callahan, Jeff Douglas, John Durso, Andrew Ehrenberg, Molson Export, Lawrence Frase,

Martin Gilchrist, Charles Lewis, Alan MacEachren, John Mazzeo, Bob Mislevy, Carol Myford, Jim Ramsay, John Rolph, Mary Vaiana, Steven Wang, Warren Willingham, and Denise Woerner. I would also like to thank Patricia Clare Haskell for advice of many sorts; also for letting me keep in most of the semicolons.

There are many staff members at Springer-Verlag who have had an important role in bringing this book to market in its current form. Principal among them are Steven Pisano (production) whose good humor and sharp eye were instrumental in producing a work that was as good as he could make it; Karen Philips (designer) who not only produced the cover but who also took seriously my sometime amateurish suggestions; and Martin Gilchrist (editor) whose enthusiasm for the project and the subject brought me to Springer in the first place.

Last, it is important to note the genesis of much of this book's contents. Chapter 1 is an expansion of a 1984 article with the same title that appeared in the *American Statistician*. Chapters 2 through 20 are variations on columns I wrote in the statistics magazine *Chance* over the time period 1990 through 1996.

Howard Wainer
Princeton, New Jersey
April, 1997

Introduction

Effectively conveying information in words is a difficult task. When the information is complex, the task often becomes insuperable. Attempts to overcome this difficulty using nonlinguistic means date back to preclassical antiquity.* The continuing search for a more complete answer took a positive turn around 3800 B.C. when ancient Egyptian geographers transformed spatial information into spatial diagrams.† Since there was no change in the metaphor of display, this was not a huge intellectual leap; but imagine how much worse off we would be if they had instead transformed the same spatial information into words.

A real breakthrough in conveying quantitative information took place in eighteenth-century Britain, when William Playfair (1759–1823), the ne'er-do-well younger brother of the well-known scientist John Playfair** and a draftsman employed by James Watt, extended the graphical metaphor to nonspatial data. He invented many of the currently popular graphical forms, improved the few that already existed, and broadly popularized the idea of graphic depiction of quantitative information. Before Playfair, the use of statistical graphics was narrowly employed and even more narrowly circulated (see chapter 2). But afterwards graphs popped up everywhere, being used to convey economic information, but also for such varied purposes as showing the distribution of ignorance in England and the frequency of improvident marriages in Wales.[1]

By the end of the nineteenth century, data-based graphics were widely used, and most of the graphical arrows that fill the modern designer's quiver had been invented. Indeed, the nineteenth century is an

*Paleolithic cave art provides an early and very striking example of graphic display. Some Ice Age bone carvings of animals are intermixed with patterns of dots and strokes that archeologists have interpreted as a lunar notation system related to the animals' seasonal appearance. These are almost identical in structure, as well as degree of detail, to the engraving on the hull of the Pioneer 10 Spacecraft, which shows a drawing of a man and a woman along with a simple plotting of the Earth's location by dotted pulsar beams.

†Spatial diagrams are of course maps. The earliest maps to survive were from Mesopotamia and were developed to keep track of land shifted by river floods.

**John Playfair's activities were remarkably varied: minister, geologist, mathematician, and professor of natural philosophy at Edinburgh University. In fact, in 1805 William thanked his brother for the idea of using "lines applied to matters of finance" that William used in his 1786 book. We can only speculate why it took him nineteen years to give his brother some credit.

important contributor to the pantheon of the most wonderful graphics ever produced. But the twentieth century has also made important contributions to the theory and practice of graphical display. These contributions have been driven by the critical need for decision-makers to be able to absorb huge amounts of information. Such a need was explicitly stated by Playfair in 1786 as the driving force behind his inventions, but in the intervening two centuries, the growth of data-gathering has been extraordinary. Associated with an abundance of data is the ubiquity of the computer to manipulate those data and various kinds of output devices that can display them. Pen nibs and accounting books have been replaced by color laser printers and CD-ROMs. Static plots are now augmented by dynamic multicolored displays that were impractical without modern high-speed computing.

The widespread understanding of how to display information effectively has lagged behind the technology for gathering, manipulating, and displaying that information. Happily, there have been a number of truly exemplary discussions of data displays,[2] which for the most part have been scholarly, serious, precise, thorough, and, if the truth be told, a little dull. The dullness is surprising since the design of charts and diagrams, representing, as it does, a synthesis of art and science, has the potential for carrying the glory of each. With the notable exception of Edward Tufte's remarkable work[3] there have been too few celebrations of that glory. *Visual Revelations* is my attempt to help remedy that lack.

In the subtitle of this book is the word *deception,* which deserves some explanation. When we see a chart or diagram, we generally interpret its appearance as a sincere desire on the part of the author to inform. In the face of this sincerity, the misuse of graphical material is a perversion of communication, equivalent to putting up a detour sign that leads to an abyss. The first section of this book examines the character of such immoral behavior. The seriousness of the sin committed by chartists who deceive is multiplied when it is viewed in the context of how powerful a tool for communication graphical methods can be. In section II, on the other hand, are reproduced some very wonderful successes. Napoleon Bonaparte appears in this section (chapter 4), although Ross Perot doesn't show up until much later (chapter 15). Between the two, you will meet some unfamiliar characters like the nineteenth-century French straphanger E.J. Marey, who provides help to anyone who has ever tried to figure out a train schedule, as well as familiar figures like Florence Nightingale, who appears here in an unfamiliar role.

This book can be read as a series of episodes; one can jump in and read any chapter, and it should, more or less, stand on its own, but the linear order of their presentation does add a little bit of useful structure. There is no ponderous theoretical structure that underlies the advice given. But there is a coherent message: revelation accompanies simplicity. This should not be confused to mean that successful displays must be very limited in content. Actually, quite the opposite is true. We must always measure the grace and simplicity of the display relative to the complexity of its message. Thus, when two charts contain the same information, the simpler will also be the more revealing. Graphic display achieves its highest goals when it allows access to important, complex information; when, in John Tukey's words, "it forces us to see what we never expected."

It is my hope that after traveling down the path of these visual revelations, your standards will change: your appreciation of the elegance of a good display will be increased, your tolerance for poor displays will be decreased, and your ability to tell the difference will be sharpened. I also hope that you will enjoy the journey.

Graphical Failures

"*Our forefathers,*" wrote Shakespeare, in *Henry VI, Part II* (Act IV), "*had no other books but the score and the tally.*" But both score and tally represented an advance from more literally iconic predecessors to the concept of abstract number. The recently discovered Combe d'Arc cave drawings in France indicated that ancient shepherds kept track of the size of their flocks through accurate drawings of them. This work was sufficiently tedious so that it was replaced by a simplified, idealized form. This simplification was later replaced by Shakespeare's tally (//////) and eventually by the abstract form "6 sheep." The movement from literal to abstract representation of information traces the history of graphic display. Note that one result of this movement is that our use of the term *graphic* has come to mean the opposite of *graphic* in the sense of literal lifelikeness.

Human beings have been using iconic representations such as maps for millennia. The spatial representation of space is an obvious metaphor. But the spatial representation of nonspatial data—such as showing rising and falling imports over time as a rising and falling line—is both more recent and a deeper intellectual accomplishment.

This major conceptual breakthrough in graphical presentation came in 1786 with the publication of William Playfair's *Political Atlas*, in which spatial dimensions were used to represent nonspatial, quantitative, idiographic, empirical data. Such a representation now seems natural, but before that time it was rarely done and was hence quite an accomplishment. Notably, in addition to the statistical line chart, Playfair at one fell swoop single-handedly invented most of the remaining forms of the statistical graphic repertoire used today—the bar chart and histogram, the surface chart, and the circle diagram, or "pie chart."

I commemorated the bicentennial anniversary of the publication of Playfair's *Atlas* by visiting the corporate offices of a tobacco company (now a tobacco and baked goods company) to give a talk on effective data display. In preparation I searched for an effective display on a topic that I thought would be of interest to them. I found one in the Surgeon General's Report on Smoking.[1]

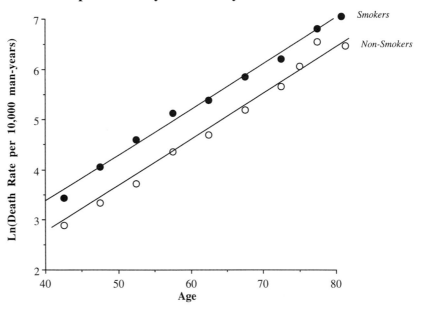

**Death Rate (Log Scale) Plotted Against Age
Prospective Study of Mortality in U.S. Veterans**

FIGURE 1. The mortality rates for smokers and nonsmokers at various ages shown in a log scale.

In figure 1 we see clearly that no matter how old you are, you are likely to die sooner if you smoke. I thought it effectively summarized a great deal of information. My audience unanimously disagreed.

Later that afternoon, as I spent some time with my host discussing these surprising objections, I began to understand. I sketched what I thought they might consider to be a more effective alternative (figure 2). By replacing two lines with stacked three-dimensional bars, the observation of smoking's danger, so clear in the Surgeon General's graph, becomes considerably more difficult.

"Much better," my host said, "but the dark bars that represent smokers are still visibly longer than the white ones. See if you can do better."

I tried some other alternatives, including labeling the plot in Greek (figure 3), which drew an appreciative clap on the shoulder and the encouragement to keep trying.

Smoking and Death Rates Shown By Age

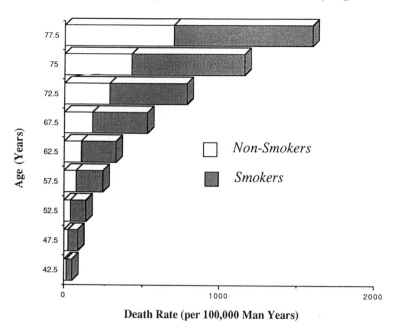

FIGURE 2. The Surgeon General's smoking data from figure 1 shown as stacked, three-dimensional bars.

Smoking and Death Rates Shown By Age

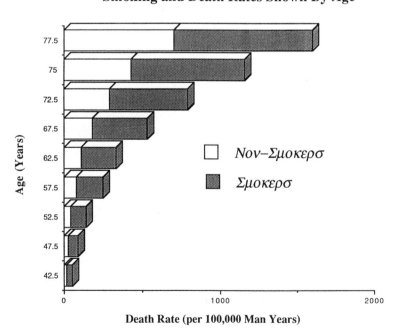

FIGURE 3. Figure 2 relabeled in Greek.

Then I hit on the answer. A double Y-axis graph! This plot (using the term with its most pejorative connotation, see chapter 8) has two lines on the same data field, but refers them to separate vertical axes.

This format is an option on every computer graphics package. Replotting the Surgeon General's data in this format yields the result shown in figure 4 with an appropriate caption. My hosts were delighted. "Now that's an effective display!" they exclaimed. Indeed.

FIGURE 4. The Surgeon General's smoking data from figure 1 shown as a double Y-axis plot and suitably relabeled.

This was the beginning of wisdom. Since Playfair, there have been many others who have attempted to advance, illustrate, and codify standards of good graphical practice. I will go into these efforts in more detail in later chapters. For now, however, I would like to discuss a much older tradition—methods of data display that instead of providing access to the complex, complicate the simple. Although such techniques are both ancient and broadly practiced, to my knowledge they have not as yet been gathered into a single source, nor carefully categorized. I believe that such a synthesis is important.

Very bad graphics surround us. They are so ubiquitous that we have become inured to their malignancy. We tend to blame ourselves when we are confused by a poorly designed graphic. Although we are often misled by graphs that distort the facts, it is only rarely that we trace the source of our misunderstanding to the offending chart. We can protect ourselves from such an assault on the truthful conveyance of information by understanding the design tricks used to construct bad displays. Just as art students attempt to learn the techniques of the old masters by studying their works and attempting to copy them, we shall try to gain a deeper understanding of the techniques of bad display through a thorough study of them. Chapter 1 is the beginning of such a study; it is built around a synthesis of the rules implicit in the design and construction of bad graphics. Chapter 2 goes further in providing the details surrounding three especially notable examples.

A Note on the Sources of Examples

The first chapter is an exposé of graphical mistakes—graphs that, perhaps out of ignorance, leave the viewer confused, misled, or uninformed. Because there is such an abundance of available error, I had to choose judiciously. It seemed unfair to pick examples from small-town newspapers, whose limited resources are apt to preclude having anyone available with both expertise and time to examine all graphics critically. Instead, I chose examples principally from the *Washington Post,* the *New York Times,* and a book written and published by the U.S. Bureau of the Census, *Social Indicators III.*

I am an admirer and daily reader of the *Times.* I suffer pangs of withdrawal if I am forced to miss it for a day. Because of my affection, I am more upset when errors appear in the *Times* than when they appear in some less-beloved vehicle. During my years in Washington, the *Washington Post* took the *Times*'s place on my breakfast table (if not in my heart). It was during this stretch that the examples from the *Post* were culled.

"Closing averages on the human scene were mixed today. Brotherly love was down two points, while enlightened self-interest gained a half. Vanity showed no movement, and guarded optimism slipped a point in sluggish trading. Over all, the status quo remained unchanged."

FIGURE 5. *New Yorker* cartoon from January 19, 1976.

Social Indicators III is not likely to be on all readers' bedstands, although it is a worthy document. It is the third in a series of triennial reports whose goal is to present, and thus disseminate, statistical data on a variety of variables that track the social fabric of our civilization, much as economic indicators track economic trends. It is part of a much larger international social-indicators movement. Similar volumes with similar aims are produced by the United Nations and by more than thirty industrialized and developing nations. Together they provide a network of information for seeing trends and making comparisons. *Social Indicators III* is meant to provide the material for the sort of newscast envisioned by Dana Farber is his remarkably apt cartoon (figure 5), and is, in general, so well done that when flaws occur they stick out. With their inclusion here, I hope they stick out far enough to be lopped off (or amended) in future volumes.

CHAPTER 1 How to Display Data Badly

Of good graphs it may be said what Mark Van Doren observed about brilliant conversationalists: *"In their presence others speak well."* A good graph is quiet and lets the data tell their story clearly and completely.

FIGURE 1. *New Yorker* cartoon from April 4, 1983.

Why is the hero of figure 1 being rewarded? Because he drew a good graph? No. Because he has good data. When looking at a good graph, your response should never be "what a great graph!" but "what interesting data!" A graph that calls attention to itself pictorially is almost surely a failure.

We shall begin our synthesis of methods for badly displaying data with a detour through the means and goals of good display. I do this in the firm belief that without recognizing the bounds of good practice, we might accidentally produce a display of some legitimate value.

**THE AIM OF GOOD DATA GRAPHICS IS TO
DISPLAY DATA ACCURATELY AND CLEARLY.**

Let us use this definition for categorizing methods of bad data display. The definition has three parts: showing data, showing data accurately, showing data clearly.

Thus, if we wish to display data badly, we have three avenues to follow.

A. Don't show much data.
B. Show the data inaccurately.
C. Obfuscate the data.

Let us examine them in sequence, parse them into some of their components, and see whether we can identify means for measuring the success of each strategy.

Don't Show Data

Obviously, if the aim of a good display includes conveying information, the less information carried in the display the worse it is. Edward Tufte[1] has converted this notion into the *data density index*, defined as *"the number of numbers plotted per square inch,"* an easily calculated index that is often surprisingly informative. In popular and technical media we have found a range from 0.1 to 362, with the data graphics in *Pravda* occupying the very bottom of the list—a model for empty display.

This gives us the first principle of bad data display:

RULE 1   Show as little data as possible
            (minimize the data density).

What does a data graphic with a data density of 0.1 look like? Figure 2 is a graphic from the Bureau of the Census's book *Social Indicators III*. The graphic's original size was seven inches by nine inches. It contains nine numbers, one for each of the nine countries shown, spread over sixty-three square inches of page. Thus its data density is calculated as $9/63$, or about 0.1, numbers per square inch. The median data graph in *Social Indicators III* has a data density of 0.6 numbers per square inch, so this one is not a particularly unusual choice.

Shown in figure 3 is a plot from the *Journal of the American Statistical Association*[2] with a data density of 0.5 (it shows four numbers in eight square inches). This is unusual for *JASA*, where the median data graph has a density of 27. In defense of the producers of this plot, the point of the graph is to show that a method of analysis suggested by a critic of their paper was not fruitful. I suspect that prose would have worked as well. This graph, while succeeding in conveying almost no information, is a failure in the pantheon of poor graphics because no one is fooled into thinking that anything is there. The trick is to

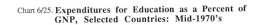

Chart 6/25. **Expenditures for Education as a Percent of GNP, Selected Countries: Mid-1970's**

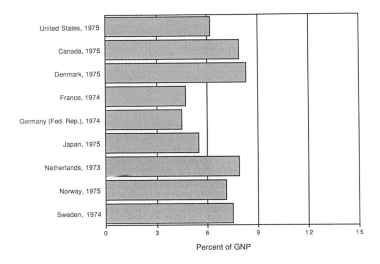

Data Density = 9 numbers /63 sq. ins. = .14

FIGURE 2. Chart 6/25 from *Social Indicators III* showing expenditures for education for nine countries as a function of GNP.

simultaneously convey no information while hiding that fact from the viewer, which is no trick for most successful politicians.

While arguments can be made that high data density does not imply that a graphic is good, nor low density that it is bad, it does reflect on the efficiency of the transmission of information. Obviously, if we hold clarity and accuracy constant, more information is better than

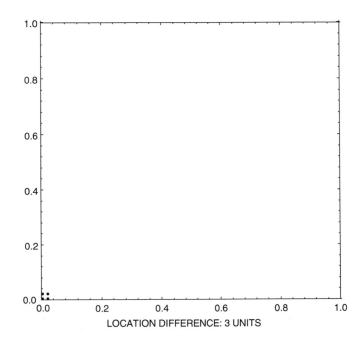

FIGURE 3. A graph of obviously low data density.

less. One of the great assets of graphic techniques is that they can convey large amounts of information in a small space. The only publication we have found that regularly has lower information density graphics than *Social Indicators III* is a typical issue of *Pravda*. Shown in figure 4 are some summary values for the data density of graphics in a variety of publications.

Median Data Densities for the Graphics in Some Well-Known Publications

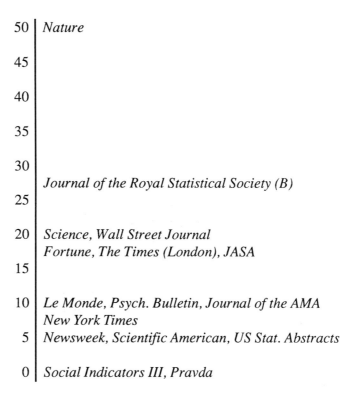

FIGURE 4. A stem and leaf diagram showing the median data densities for sixteen publications. (Reprinted with permission from Tufte, 1983.)

50	*Nature*
45	
40	
35	
30	
25	*Journal of the Royal Statistical Society (B)*
20	*Science, Wall Street Journal*
	Fortune, The Times (London), JASA
15	
10	*Le Monde, Psych. Bulletin, Journal of the AMA*
	New York Times
5	*Newsweek, Scientific American, US Stat. Abstracts*
0	*Social Indicators III, Pravda*

Higher density graphs are easily found. Note in figure 5 a plot of weather from the *New York Times* (January 3, 1997) that contains 1,849 numbers in 34 square inches for a data density of 54. Yet, despite this high density of information, it is comprehensible. We can see both the large (it is warmer in the summer than the winter) and the small (on the hottest day in the year the temperature reached 96°). We see the usual (the 365 normal highs and lows) and the unusual (during most of the last two weeks of February the temperature was in the mid-50s). We can also make interesting comparisons (December 20th was considerably warmer than April 15th!). This graph demonstrates that a graph whose data density reaches into the fifties does not preclude it from being an evocative, informative display.

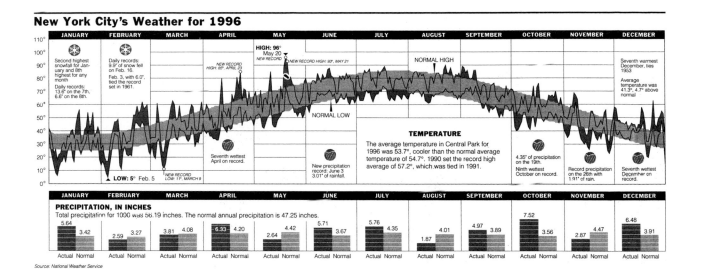

New York City's Weather for 1996

Source: National Weather Service

FIGURE 5. A year-end weather summary graph from the *New York Times*.

We noted earlier that when a graph contains little or no information the plot can look empty (figure 3) and thus raise the viewer's suspicions that nothing is to be communicated. A way to allay these suspicions is to fill up the plot with nondata figurations, what Edward Tufte has termed "Chartjunk." In figure 6 is a plot of the labor productivity of Japan relative to that of the United States. It contains one number for each of three years. Obviously, a graph of such sparse information would have a lot of blank space, but by placing the evocative Rising Sun out front, both Old Glory and the paucity of information in the graph are hidden from the reader.

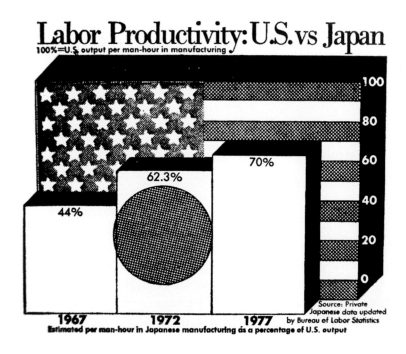

FIGURE 6. A graph with low data density filled in with chartjunk from the *Washington Post*, 1978.

Chartjunk can take many forms. Some of these are depicted in figure 7. The more of these nondata figurations included in a graph, the smaller the visual effect of the data. Edward Tufte devised a convenient measure of this practice, the "data/ink ratio": the ratio of the amount of ink used to graph the data divided by the total amount of ink in the graph. The closer to zero this ratio gets, the worse the graph. Obviously, the greater the proportion of ink used for chartjunk the lower the data/ink ratio.

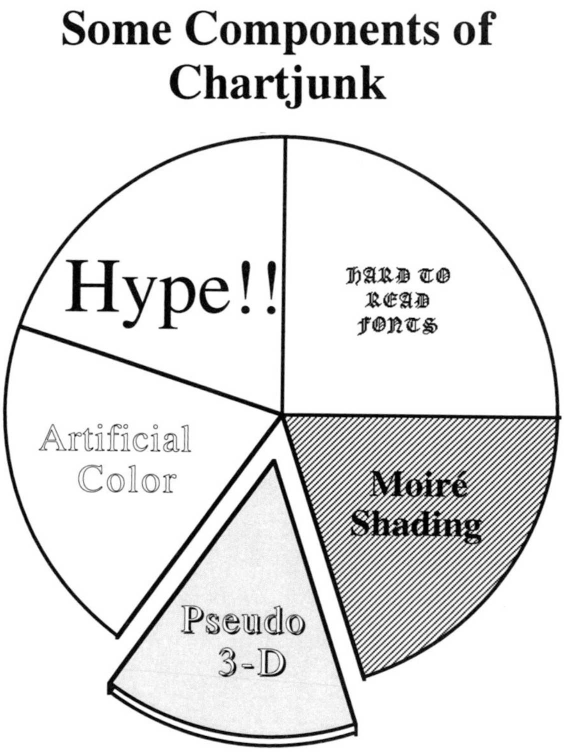

Some Components of Chartjunk

FIGURE 7. Some components of chartjunk baked in a pie.

The notion of the data/ink ratio brings us to the second principle of bad data display.

RULE 2 Hide what data you do show
(minimize the data/ink ratio).

Hiding the data in the grid

The grid is useful for plotting the points by hand, but once they are plotted, the grid rarely serves any further purpose. Thus, to display data badly, use a fine grid and plot the points dimly. This is exemplified by a plot by three political scientists[3] intended to show the relationship between actual voter registration rates and those predicted by their mathematical model. It is accurately reproduced in figure 8. Because of the amount of ink used up by the grid, the data/ink ratio for this plot is close to zero.

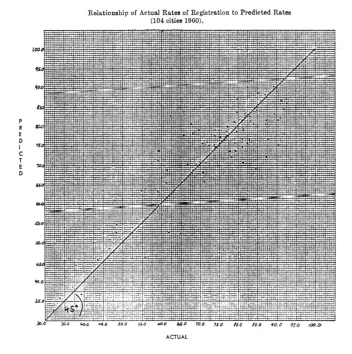

FIGURE 8. Hiding the data in the grid.

Can it get any worse? The paper from which this plot was abstracted was subsequently anthologized,[4] and the editor, William J. Crotty, hit upon a plan to decrease the data-ink ratio still further. He omitted the data entirely, emphasized the grid, and still maintained the extraneous 105% border. He utilized a full page of a book to reproduce a piece of graph paper with a diagonal line (figure 9). This is a realization of the ultimate in empty display—a zero data density and zero data/ink ratio!

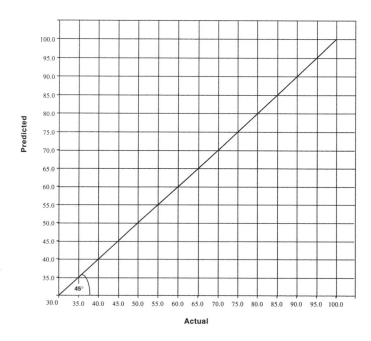

FIGURE 9. The ultimate—a completely contentless graph.

Tufte,[5] in an almost visual subtraction of Crotty's plot from the original, produced a graph with a data/ink ratio of about 0.7 (see figure 10). Tufte's display actually allows the viewer to judge the extent to which the model is a reasonable predictor of the data.

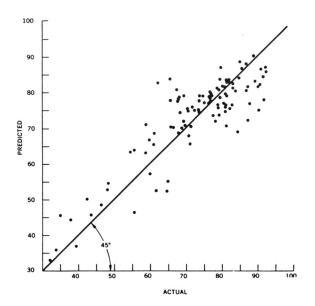

FIGURE 10. A redone example of the data from figure 8.

Graphic display has much in common with architecture. A building must look nice, but it must also keep the rain out, and its toilets must flush. A graph must also look nice, but it too must work, which is why it is often useful to view graphical situations within an architectural metaphor. This is one of those times. The grid on which the graph is plotted is like the scaffolding surrounding a building under construction. But when the building is complete, the scaffolding is removed. So too should it be with the graphical grid. Leaving the grid in detracts both from a graph's beauty and its ability to communicate. Of course, grids are now largely anachronistic since modern, computer-generated graphs don't need them.

Hiding the data in the scale

The process of hiding the data in the scale consists of blowing up the scale (thereby looking at the data from a distance) so that the magnitude of the scale obscures any variation in the data. One justification of this practice is that "honesty requires that we start the scale at zero." And there are other sorts of sophistry.

In figure 11 is a plot (from *Social Indicators III*) that effectively hides the growth of private schools in the scale.

A redrawing of the number of private elementary schools on a different scale conveys the massive growth that took place during the

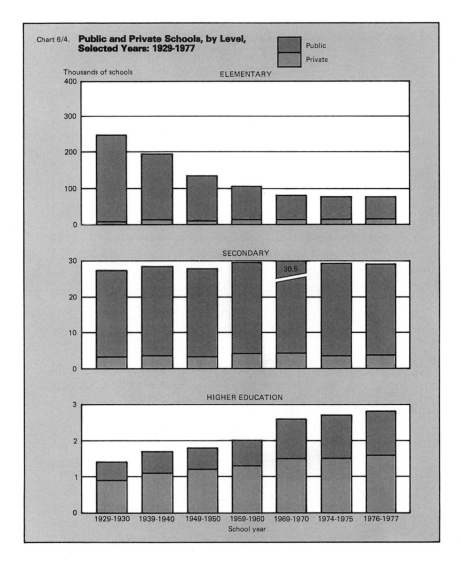

FIGURE 11. Hiding the data in the scale. (See insert for color version.)

mid-1950s (figure 12). One plausible explanation for this previously invisible increase in private elementary schools is the reaction to the *Brown v. Board of Education of Topeka Kansas* Supreme Court decision in 1954. Of course this might not be related, but even the suggestion was missing with the original choice of scale.

Another example of scale manipulation originated in the Reagan White House* and is shown as part of the discussion surrounding rule 5.

We have seen that we can display data badly by not including them (rule 1) or by hiding them (rule 2). We can measure the extent to which we are successful in excluding the data through the data density. We can sometimes fool viewers into thinking that we have included enough data to be informative through the incorporation of chartjunk. We can hide the data by using an overabundance of chartjunk or by cleverly choosing the scale so that the data disappear. A measure of the success we have achieved in hiding the data is through the data/ink ratio.

*Graphics in politics occur less frequently than one might suspect. Graphs tend to be too explicit to allow the sorts of vagueness favored by successful politicians. This issue is explored more fully in chapter 15.

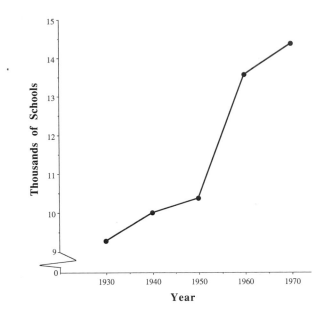

FIGURE 12. Expanding the scale and showing the data for the number of private elementary schools from figure 11.

Show Data Inaccurately

The essence of a graphical display is that a set of numbers, ordered by their size, are then represented by an appropriate visual metaphor—the magnitude and order of the metaphorical representation matches the numbers. By ignoring or distorting this concept we can display data badly.

RULE 3   Ignore the visual metaphor altogether.

When the data are ordered and when the visual metaphor also has a natural order, if you shuffle the relationship a bad display will surely emerge. In figure 13, note that the bar labeled "14.1" is longer than the bar labeled "18," probably an error made by the artist in an attempt to comfortably fit the four-character-long "14.1" into a bar. Of course, rounding it to "14" would have worked very well indeed. Such thoughtless errors are called *graphos* (the analogue of textual *typos*).

FIGURE 13. Ignoring the visual metaphor by letting a longer bar segment represent a smaller amount of coal (from the *New York Times*, 1978).

A New Set of Projections for the U.S. Supply of Energy

Compared are two projections of United States energy supply in the year 2000 made by the President's Council of Envirnonental Quality and the actual 1977 supply. All figures are in "quads," units of measurement that represent a million billion — one quadrillion — British thermal units (B.T.U.'s), a standard measure of energy.

Oil and gas   Solar*

Nuclear   Coal

1977   Total 77.5   56.5   4.2   14.1   2.7

2000 — A
Emphasizes energy conservation   Total 85   40   19   8   18

2000 — B
Emphasizes increased energy production   Total 120   46   19   18   37

* Includes all renewable energy sources   Source: President's Council on Environmental Quality

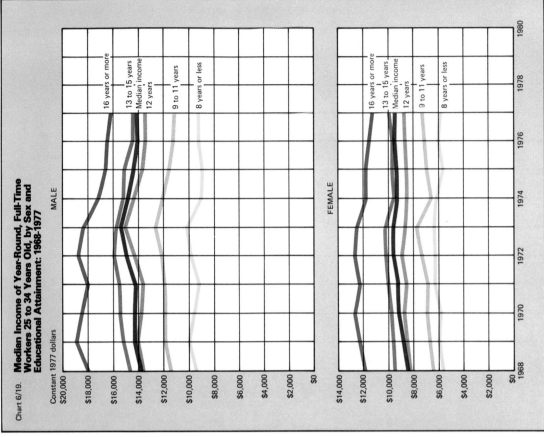

CHAPTER 1, FIGURE 27. Emphasizing the trivial: Hiding the main effect of sex differences in income through the vertical placement of plots.

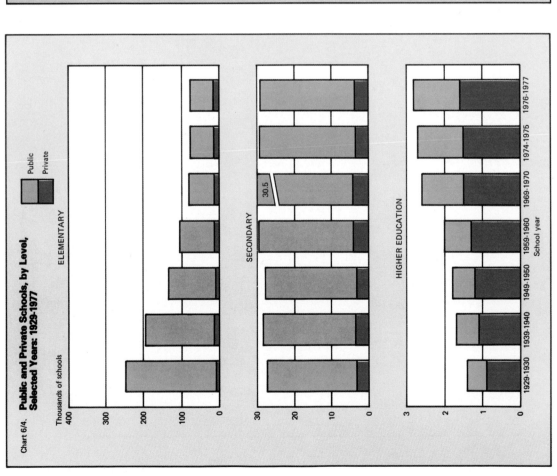

CHAPTER 1, FIGURE 11. Hiding the data in the scale.

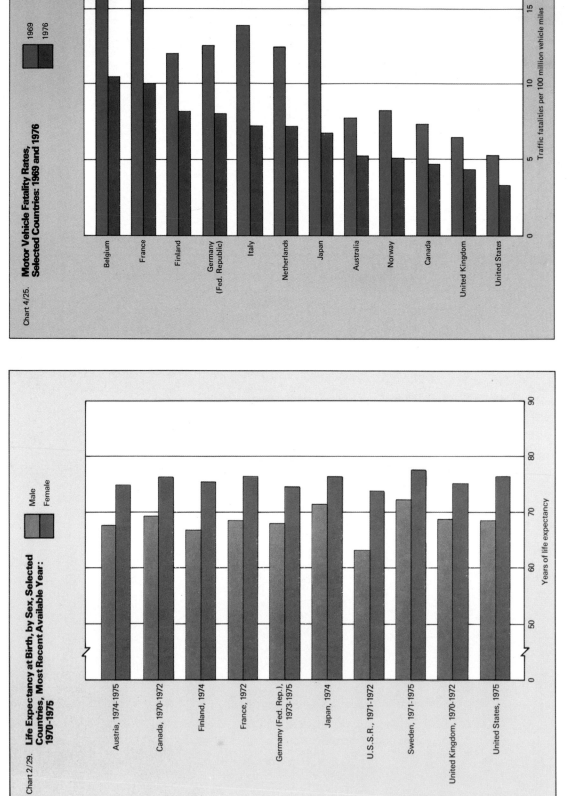

Chart 4/25. **Motor Vehicle Fatality Rates, Selected Countries: 1969 and 1976**

1969
1976

Belgium
France
Finland
Germany (Fed. Republic)
Italy
Netherlands
Japan
Australia
Norway
Canada
United Kingdom
United States

Traffic fatalities per 100 million vehicle miles

CHAPTER 1, FIGURE 37. Ordering the bar chart by the data tells the tale a bit more clearly.

Chart 2/29. **Life Expectancy at Birth, by Sex, Selected Countries, Most Recent Available Year: 1970-1975**

Male
Female

Austria, 1974-1975
Canada, 1970-1972
Finland, 1974
France, 1972
Germany (Fed. Rep.), 1973-1975
Japan, 1974
U.S.S.R., 1971-1972
Sweden, 1971-1975
United Kingdom, 1970-1972
United States, 1975

Years of life expectancy

CHAPTER 1, FIGURE 35. Austria first! Obscuring the data structure in some life expectancy data by alphabetizing the plot.

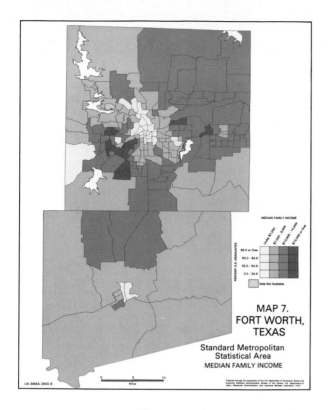

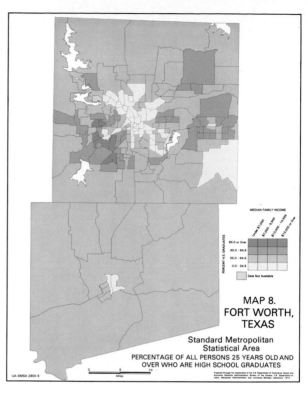

CHAPTER 1, FIGURE 44. The geographic distribution of median family income in Fort Worth, Texas, in 1974.

CHAPTER 1, FIGURE 45. The geographic distribution of percentage of high-school graduates in Fort Worth, Texas, in 1974.

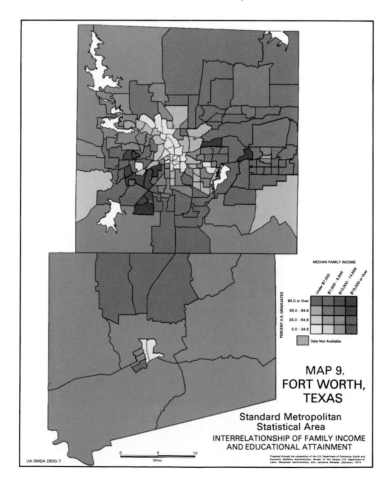

CHAPTER 1, FIGURE 46. The geographic distribution of both median family income and percentage of high-school graduates in Fort Worth, Texas, in 1974, shown as a two-variable color map.

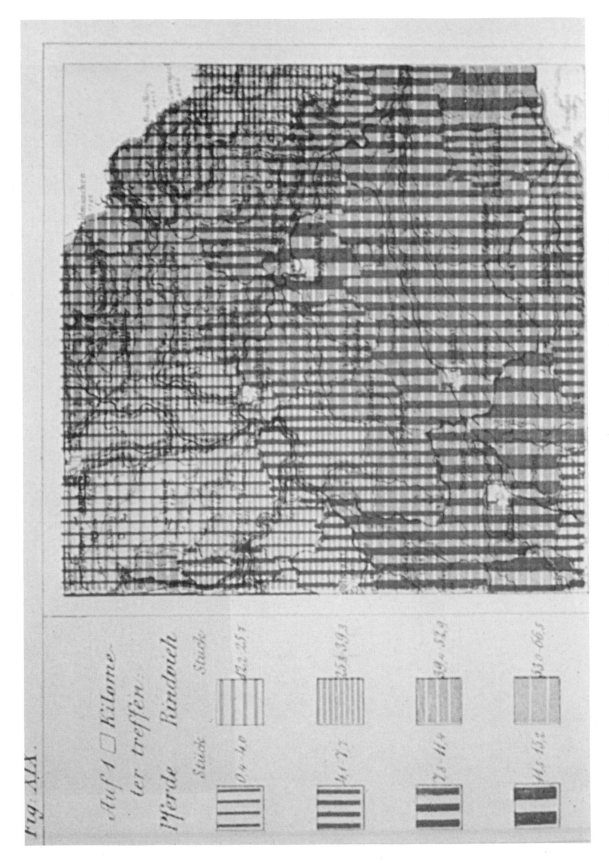

CHAPTER 1, FIGURE 47. A two-variable color map showing the joint distribution of horses (*Pferde*) and cattle (*Rindvieh*) in eastern Bavaria in 1874.

U.S. trade with China and Taiwan

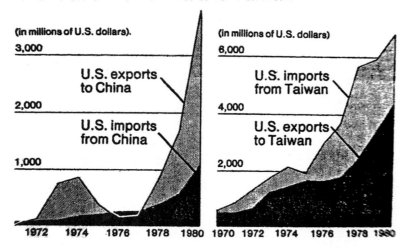

FIGURE 14. Reversing the metaphor in mid-graph while changing scales on both axes (from the *New York Times*, June 14, 1981).

Another method, driven by what often seem to be more sinister motives, is to change the meaning of the metaphor in the middle of the plot.

In the left panel of figure 14 the dark shading, which represents imports *from* China, lies below the line corresponding to exports *to* China. This configuration represents a positive balance of trade for the United States with China. A quick glance at the right panel shows what appears to be the same happy situation with Taiwan. A closer look indicates that this is not the case. The metaphor is reversed; shading is now used for exports. The distortion is compounded since the scale of the right panel is double that of the left. Thus not only do we have a negative balance of trade with Taiwan, but it is twice as large as it first appears. There is also a difference in the time scale, but that is minor. A redone version is shown in figure 15.

FIGURE 15. Figure 14 redone with a consistent scale and visual metaphor.

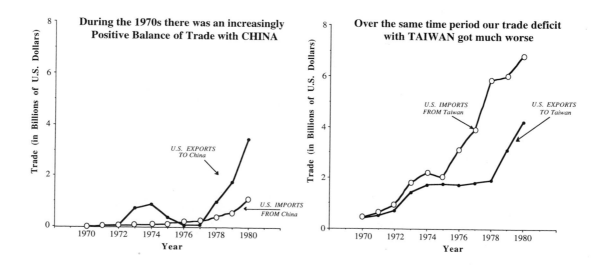

FIGURE 16. A plot on the same topic done well two centuries earlier.

The misuse of graphics in the depiction of trade balances is especially ironic since wonderful depictions of such data formed a common theme in Playfair's work. In figure 16 is a two-hundred-year-old graph that tells the story. Two such plots would have illustrated the story surrounding U.S. and Taiwanese trade quite clearly (this is another example of rule 12). Figure 16 has another feature of interest, which I will discuss as part of rule 8.

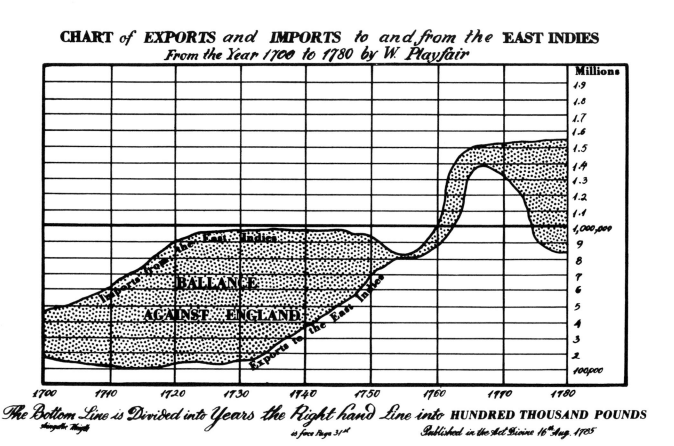

CHART of EXPORTS and IMPORTS to and from the EAST INDIES
From the Year 1700 to 1780 by W. Playfair

The Bottom Line is Divided into Years the Right hand Line into HUNDRED THOUSAND POUNDS

RULE 4   Only order matters.

One trick often utilized is to use length as the visual metaphor when area is what is perceived. This was used quite effectively by the *Washington Post* in figure 17. Note that this graph also has a low data density (0.1), and its data-ink ratio is close to zero. If we replot this with less pizzazz, but more honesty, we get a clearer idea of the size of inflation as well as seeing that most of it took place during the Nixon-Ford administrations (figure 18).

Using these two plots allows us to calculate a measure of the distortion. One such measure is Tufte's "lie factor," which in this instance is the perceived change in the value of the dollar from Eisenhower to Carter divided by the actual change. I read and measure thus:

$$\underset{\text{Actual}}{\frac{1.00 - .44}{.44} = 1.27} \qquad \underset{\text{Measured}}{\frac{22.00 - 2.06}{2.06} = 9.68}$$

$$\text{lie factor} = 9.68/1.27 = 7.62.$$

This distortion of more than 600% is substantial, but by no means a record.

Our understanding of perceptual distortions like these grew out of the experiments and formal analyses conducted by gestalt psychologists. Much of this work was done in the late nineteenth century in Wundt's laboratory in Leipzig, where this particular distortion was sufficiently well known to have its own name. Roughly translated from the German, it is called "the old goosing up the effect by squaring the eyeball trick."[6]

FIGURE 17. An example of how to goose up the effect by squaring the eyeball.

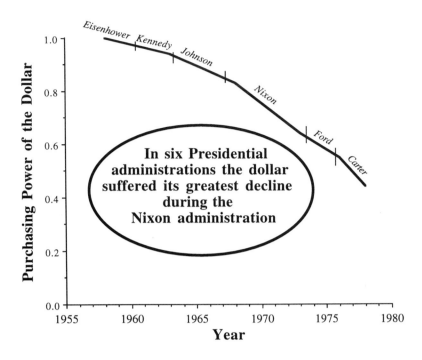

FIGURE 18. The data in figure 17 as an unadorned line chart (from Wainer, 1980).

Even greater distortion can be accomplished by using volumes rather than areas and by playing around with the scale and origin. In figure 19 the volumetric increase in beer sales is a whopping 38,200%, whereas the actual increase is a respectable 29%, for a lie factor of 131,724%. A more accurate depiction is shown in figure 20.

In discussing these kinds of distortions, people who prepare graphics for the media (mostly newspapers) had two excuses. They (1) "just wanted to show that the dollar was shrinking," or (2) explained that "after all, the correct amount was written just below the dollar."

FIGURE 19. Cubing the visual effect and choosing the origin to yield a near record lie factor of over 131,000% (from the *Washington Post*).

U.S. Beer Sales and Schlitz's Share

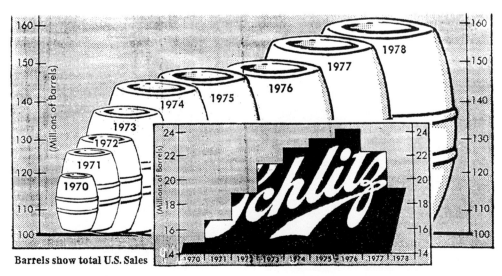

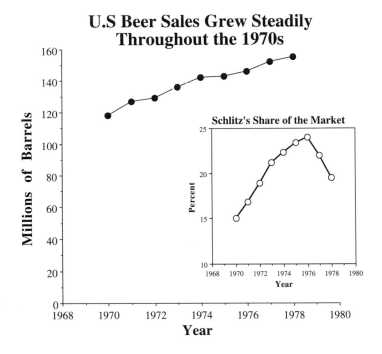

FIGURE 20. Data from figure 19 redone without tricks (from Wainer, 1980).

I guess excuse number two means that they felt that it is OK to lie in one place if you tell the truth in another. Excuse number one reflects a limited understanding of the full power of what a number represents. Numbers can be ordered, certainly; 14 is greater than 3. But the real power of numbers is in their magnitudes. Hearing that your salary is going up next year is good news, but who could hear that without wanting to ask "How much?"

RULE 5   Graph data out of context.

The value of a fact shrinks enormously without context (see chapter 16 for an expansion of this idea). Knowing that more than 600,000 Americans were killed in the Civil War is horrifying, but its impact is multiplied when you learn that it is 200,000 *more* than the number of Americans killed in World War II.

By choosing the interval displayed carefully we can often modify the perception of the graph (particularly for time series data). A precipitous drop can disappear by choosing a starting date just after the drop. In figure 21 the image we retain is of a stable unemployment rate of about six percent. But why did the graph start in January? Would our perceptions change if it started earlier? Figure 22 starts the series in October while expanding the scale to allow a closer look. We see now that there was a sharp drop in unemployment at the end of 1977. Is this too an artifact of the starting point? If we began the series in August or July would this decline change to be just a random meander? No. For by plotting the average rate for all of 1977 we see that the further back into 1977 we go, the higher the unemployment rate.

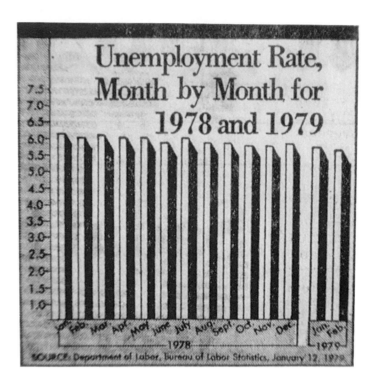

FIGURE 21. Hiding the effect by the careful choice of scale and origin (from the *Washington Post*).

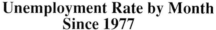

FIGURE 22. Regraph of data from figure 21 with expanded scale, different starting point, and previous year's average added for context (from Wainer, 1980).

Similarly, we can turn slight meanders into sharp changes by focusing in on a single meander and expanding the scale. Consider the comparison between two versions of President Reagan's 1981 budget plans shown in figure 23. The inset was taken from the president's television message to the American people. By expanding the scale and omitting the overall level, we take away the message that "our bill" may cost a little more than "their bill" now, but in a few years it will save us substantially. By providing context, the "neutral view,"* allows us to understand better the size of the predicted effect of the two tax proposals.

We have seen how the choice of scale, though arbitrary, can have profound effects upon the perception of the display. Automatic rules do not always work; wisdom and honesty are always required.

In this section we have discussed three rules for the accurate display of data. One can compromise accuracy by ignoring visual metaphors (rule 3), by paying attention only to the order of the numbers and not their magnitude (rule 4), or by showing data out of context (rule 5). We have advocated the use of Tufte's lie factor as a way of measuring the extent to which the accuracy of the data has been compromised by the display.

*More accurately called "the *New York Times* view."

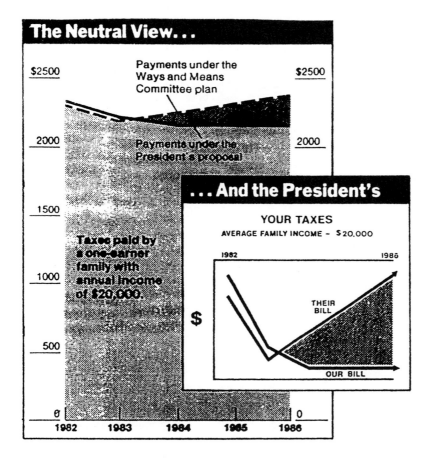

FIGURE 23. *New York Times* graphs showing how lack of context changes our perceptions about alternative tax bills.

Obfuscate the Data

In this section we introduce and illustrate six more rules for displaying data badly. Rules 6 through 11 fall broadly under the heading of how to obscure the data. These techniques are changing the scale in mid-axis, emphasizing the trivial, jiggling the base line, ordering the chart by a characteristic unrelated to the data, labeling poorly, and including more dimensions and/or decimal places than are justified or needed. Each method, alone or in combination with the others, will produce graphs and tables of little use. Usually their common effect will be to leave us uninformed about the points of interest in the data. Sometimes, however, they will misinform us. The physicians' income plot in figure 25 is a prime example. The twelfth rule is a perversion of George Santayana's observation that "those who cannot remember the past are condemned to repeat it."*

RULE 6   Change scales in mid-axis.

Changing scales in mid-axis is a powerful technique that can make large differences look small and make exponential changes look linear.

   In figure 24 is a graph that supports the associated story about the skyrocketing circulation of the *New York Post* relative to the plummeting *Daily News* circulation. The reason given: New Yorkers "trust" the *Post*.

*From *The Life of Reason, II. Reason in Society*. Santayana was paraphrasing Thucydides in his famous *History of the Peloponnesian War* (Book 1, Section 1).

FIGURE 24. Changing the scale in mid-axis to make large differences seem small (from the *New York Post*, May 12, 1981).

The soaraway Post — the daily paper New Yorkers trust

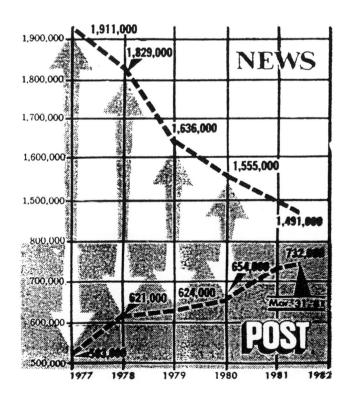

It takes a careful look to note the 700,000 jump the scale makes between the two lines. Such a bald-faced shift in scale might fool New York's naive newspaper readers, but surely a more subtle technique is required to fool Washington's sophisticates.

The *Washington Post* proved up to the task. Figure 25 is a plot of physicians' incomes from 1939 to 1976. It appears linear with a slight tapering off in later years. A careful look at the scale shows that it starts out plotting every eight years and ends up plotting each year. The message then is that physicians' incomes seem to have gone up linearly. But the years during the decade of the 1970s pass much more slowly than the years of the 1940s.* Even Washington's sophisticated inhabitants would have trouble interpreting this one correctly. A more regular scale (in figure 26), in which time is allowed to move in its inexorably linear fashion, tells quite a different story. Physicians' incomes have skyrocketed. The editorial comment indicating the date of Medicare's onset may be used to counter arguments about the effects of socialized medicine on physicians' incomes.

*The increase of about $5,000 from 1939 until 1947 looks slightly larger than the $4,300 increase between 1975 and 1976. But the earlier increase, since it spans an eight-year period, represents an annual increase of less than $700. Thus the annual growth of physicians' incomes is more than six times what it once was. But the graph makes it look as though the increases were constant.

FIGURE 25. Changing scale in mid-axis to make exponential growth linear (from the *Washington Post*, Jan. 11, 1979, in an article titled "Pay, Practices of Doctors on Examining Table" by Victor Cohn and Peter Milius).

FIGURE 26. Data from figure 25
redone with a linear scale
(from Wainer, 1980).

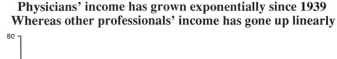

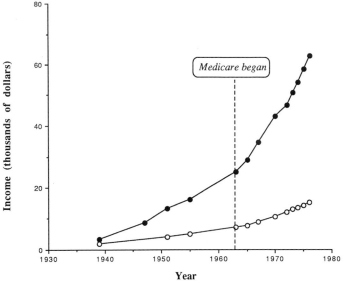

RULE 7   **Emphasize the trivial
(ignore the important).**

Sometimes the data to be displayed have some aspects that are important and others that are trivial. The graph can be made worse by emphasizing the trivial parts while hiding the important. In figure 27 we
have a page from *Social Indicators III* that compares the income levels
of men and women by educational levels. This graph reveals the un-
surprising result that better-educated individuals are paid better than
poorer-educated ones and that changes in income across time, when
shown in constant dollars, are modest. The comparison of greatest interest and current concern, comparing salaries between sexes among
people at the same level of education, must be made clumsily by vertically transposing from one graph to another. It seems clear that rule 7
must have been operating here, for it would have been easy to place
the graphs side by side and allow the comparison of interest to be
made more directly (figure 28).

   But suppose we don't know enough to specify in advance what aspect is most important. Does the structure of the data themselves
suggest a helpful display? The nature of communications media (paper, video displays, etc.) as well as our too Euclidean perceptions
means that most displays must be arrayed on a two-dimensional surface. That means that only two data dimensions are spatially represented. A third (or higher) dimension must be rendered with some
other sort of metaphor. Such a rendering must be read and not
viewed. Such "read" variables are more dimly, and less accurately, perceived. Geographic maps use the two dimensions of the display plane

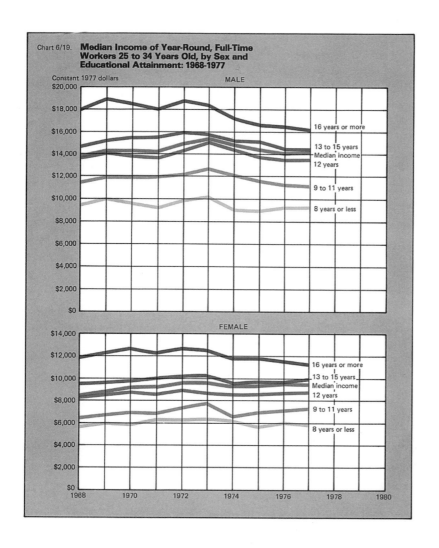

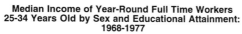

FIGURE 27. Emphasizing the trivial: Hiding the main effect of sex differences in income through the vertical placement of plots. (See insert for color version.)

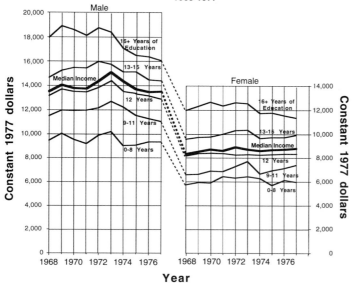

FIGURE 28. Figure 27 redone with the two plots horizontally opposed, showing the size of sex differences more clearly.

to depict north-southness and east-westness in a natural way. Their depiction of any third dimension, say altitude, is often done with contour lines. Our perception of the relative distance between places is both more accurate and accomplished more naturally than that of their relative elevations.

Viewing the earnings data from this perspective, we note that there are two variables that seem to have a large effect (education and sex) and that the changes across time are modest in comparison. A sensible display ought to use the two dimensions of the plot to show these large effects clearly and place the smallish time trend into the background. Figure 29 is an attempt to do this. The vertical bar represents the entire range of incomes over the ten years reported. We can see at least three things immediately that were not evident before.

1. The parallel nature of the curves for men and women's earnings shows that education has the same relative effect on both sexes.
2. By tracing the earnings level for women who have graduated from college over to the left, we see that it is about the same as for men with less than a high-school diploma.
3. By tracing the earnings level for men with only a grade-school education (0–8 years) over to the right, we see that it is about the same as for women with some college (13–15 years).

Viewing these data for the decade 1968–1977 provides a context for understanding at least one of the goals of the women's movement. A measure of progress would be easily seen from a parallel graph for the 20 years since then.

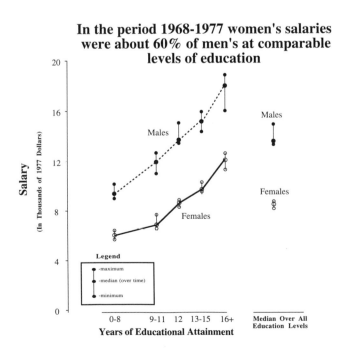

FIGURE 29. Figure 28 redone with the large effects of sex and education emphasized and the small-time trend suppressed.

RULE 8 Jiggle the baseline.

Human beings are good at making comparisons with a straight line. Graphical comparisons are thus always easier when the quantities being compared start from a common base. Conversely, we can always worsen the graph by starting from different bases or by making comparisons from a curve.

Consider the pairs of curves shown in the left panel of figure 30.[7] Try to estimate the distance between the two curves visually. Remember that the distance is measured vertically and so the upsweep at the end confuses matters. The right panel was computed from the left by subtracting the bottom curve from the top. The difference, the residual, is then plotted explicitly. No one can look at the paired curves and tell the difference between lines that are equidistant (the residual line is flat), or have some funny dip or bump.

FIGURE 30. A graphical experiment (from Cleveland and McGill, 1984). Without looking at the corresponding right panel, try to determine the difference between the two curves in the left panel.

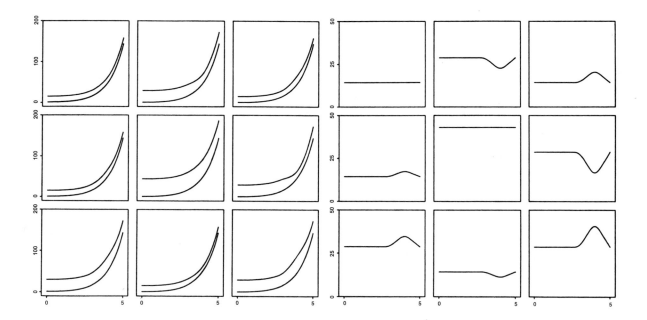

Are such effects likely to show up in the real world? Or are they merely the result of some overly pessimistic statistician? In fact, such deceptive effects are common. I will report two different graphical formats in which such an effect showed up. The first you have seen before (figure 16, reproduced here as figure 31): William Playfair's eighteenth-century depiction of England's trade with the East Indies. Consider the size of the trade imbalance in the twenty years beginning in 1750, what might be called the "Isthmus of Playfair." The visual impression is that the trade imbalance shrinks to almost nothing in 1755 and then grows slightly until 1770, when it begins to grow substantially.

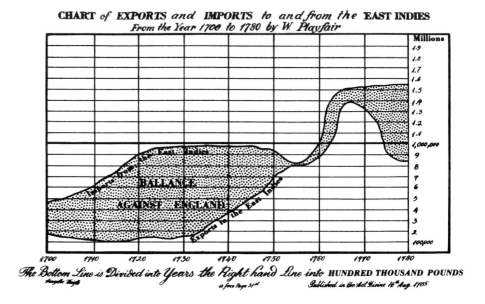

FIGURE 31. William Playfair's eighteenth-century graph of England's imports and exports with the East Indies (from Cleveland and McGill, 1984).

A different, and more correct, impression is obtained if we actually plot the difference between the two curves (figure 32). We see a substantial bump in the trade imbalance in 1762–1763, completely invisible before.

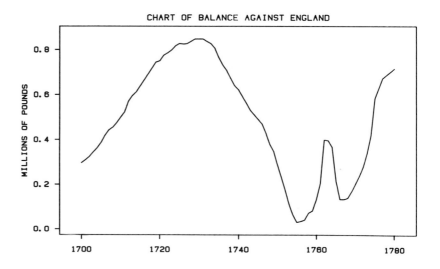

FIGURE 32. A graph of the difference between West Indies imports and exports showing explicitly the previously invisible jump in the 1760s (from Cleveland and McGill, 1984).

Jumping forward two hundred years for our next example provides us with a series of stacked bar charts (figure 33). The total length of each bar gives us the total amount of petroleum stocks in OECD (Organization for Economic Cooperation and Development) countries during the decade 1977–1986. Each bar is divided into four sections representing the stocks held in the United States, Japan, West Germany, and all other OECD countries. The bottom of the bar always starts from a flat base, and so we can see that U.S. petroleum

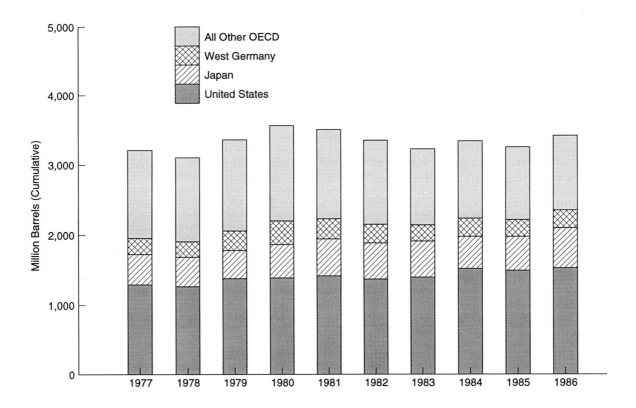

stocks have been increasing. However, judgments about the trends in the stocks of the other three components cannot be made accurately since they all start at different places. Plotting each country separately (figure 34) provides us with a more accurate view. All that is lost in this disaggregated display is a depiction the trend of all OECD countries taken together. But with one more panel in the overall display this too could have been included. What is missing in the original display is the sharp decline in petroleum stock evidenced by the aggregation of "All Other OECD." Stacked bar charts are a ubiquitous feature of all computer graphics routines, and they always foster this sort of deception. They trail only the double Y-axis graph in their importance as a weapon in the armory of all whose principal goal is the distortion of information.

RULE 9   Alabama first!

Ordering graphs and tables alphabetically can effectively obscure structure that would have been obvious had the display been ordered by some aspect of the data. One can defend oneself against criticism by pointing out that alphabetizing "aids in finding entries of interest." Of course, with lists of only modest length, such aids are unnecessary; with longer lists, the indexing schemes common in nineteenth-century statistical atlases provide easy lookup capability.

FIGURE 33. From the U.S. Department of Energy's *Annual Energy Review, 1986,* showing the changes in primary stocks of petroleum in OECD countries.

OECD PETROLEUM STOCKS HAVE STABILIZED

But Not All Countries Are Pulling Their Own Weight

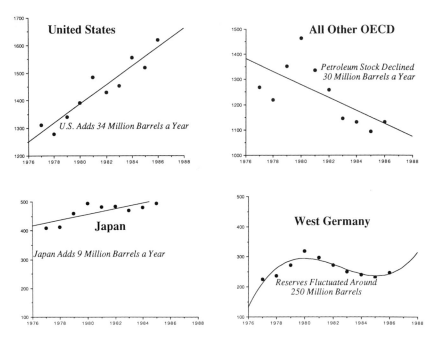

FIGURE 34. Regraphing of the data from figure 33 in which each country's data are shown relative to a straight line.

Figure 35 is another graph from *Social Indicators III*, this one showing life expectancies divided by sex for ten industrialized nations. The order of presentation is alphabetical (with the USSR positioned as Russia). The message we get is that there is little variation, and women live longer than men. Redone as a stem-and-leaf diagram (figure 36, a simple reordering of the data with spacing proportional to the numerical differences), the magnitude of the sex difference (seven years) leaps out at us. We also note that the life expectancy for men in the USSR is unusually short, whereas Soviet women seem to live about as long as women in the other countries.

Is the difference between the clarity of these two displays due, as I am suggesting, to the ordering of the countries? Or is the format of figure 36 just superior? I contend that both are true. To see this, consider figure 37, another plot from *Social Indicators III* that has the same formal structure as figure 35, except that the countries are ordered by their 1976 motor vehicle fatality rates. We easily see the profound differences (a factor of four between the U.S. and Belgium!) as well as the large advances made in auto safety over the seven years depicted. Last, the unusual gain made by Japan in auto safety is visible. I do not contend that a better display still could not be made, but the simple ordering has helped. The Japanese improvement would have been obscured by alphabetizing. In tabular presentations the effects of alphabetizing (or presenting the data ordered by some other aspect unrelated to the data) can be particularly profound (see chapter 12).

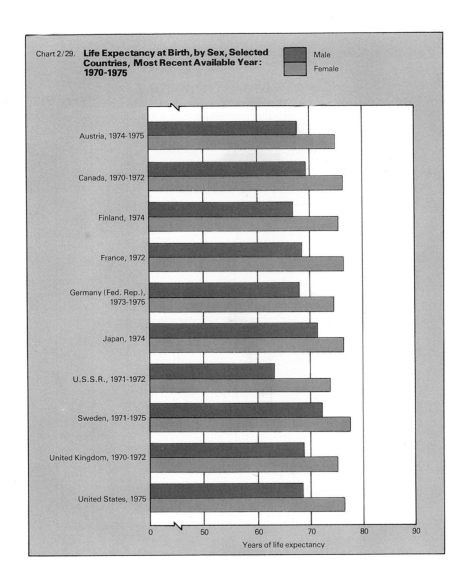

FIGURE 35. Austria first! Obscuring the data structure in some life expectancy data by alphabetizing the plot. (See insert for color version.)

Life Expectancy at Birth, By Sex
Most Recent Available Year

Women	Age	Men
Sweden	78	
France, US, Japan, Canada	77	
Finland, Austria, UK	76	
USSR, Germany	75	
	74	
	73	
	72	Sweden
	71	Japan
	70	
	69	Canada, UK, US, France
	68	Germany, Austria
	67	Finland
	66	
	65	
	64	
	63	
	62	USSR

FIGURE 36. Ordering and spacing the data from figure 35 as a stem-and-leaf diagram provides insights previously invisible.

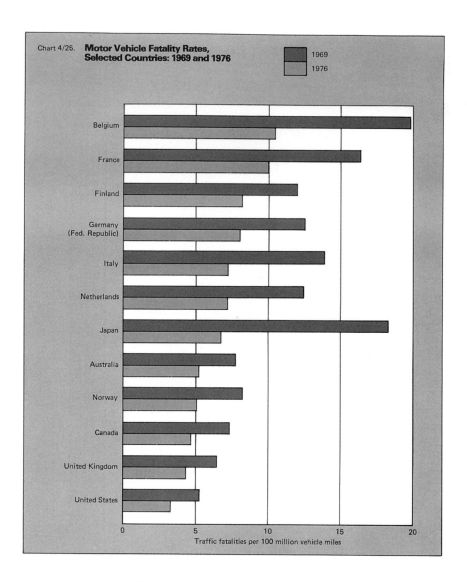

FIGURE 37. Ordering the bar chart by the data tells the tale a bit more clearly. (See insert for color version.)

RULE 10 Label: (a) illegibly, (b) incompletely, (c) incorrectly, and (d) ambiguously.

A picture may be worth a thousand words, but it may take a hundred words to make it so. Effective labels point the viewer's attention to the principal idea of the display. There are many instances of labels that either do not tell the whole story, tell the wrong story, tell two or more conflicting stories, or are so small that one cannot figure out what story they are telling. A favorite example of "small labels" is from the *New York Times* (August 1978), in which the article complains that fare cuts lower commission payments to travel agents. The graph (figure 38) supports this view until one notices the tiny label indicating that the small bar showing the decline is for just the first half of 1978. By omitting the second half of the year it omits such heavy travel periods as Labor Day, Thanksgiving, and Christmas, so that merely doubling the 1978 figure is probably not enough. Nevertheless, when this bar is doubled (figure 39), we see that relative to earlier years, the agents are doing very well indeed.

FIGURE 38. Mixing a changed metaphor with a tiny label reverses the meaning of the data.

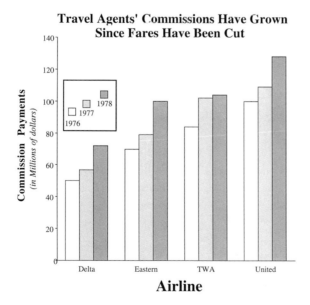

FIGURE 39. Figure 38 redrawn with 1978 data placed on a comparable basis shows that the fare cuts have been a boon to travel agents.

RULE 11 More is murkier: (a) more decimal places and (b) more dimensions.

A display can be made clearer and more truthful by presenting less (see chapter 12 for a fuller explanation). Consider the statement, "This year's school budget is $27,329,681." Who can comprehend or remember this?

If we remember anything, it is almost surely the translation, "This year's school budget is about twenty seven million dollars." Anyone who knows anything about budgets knows that little beyond the third digit has any meaning, and our memory's censoring is sensible. We often see tables in which the number of digits presented is far beyond the number that can be understood and used by a reader. These numerical data are commonly presented to more accuracy than is justified.

Table 1 below (abstracted from a much larger version in the 1962 *UN Demographic Yearbook*) is a table of life expectancies, to two decimal places. The second decimal place represents four days. Who cares about averages to this level of precision? It is not only useless information, it also makes the table harder to understand.

TABLE 1
Life Expectancy at Birth

Country	Male	Female
Argentina	56.90	61.40
Brazil	39.30	45.50
Canada	67.61	72.92
Iceland	66.10	70.30
Japan	65.37	70.26
Mexico	37.92	39.79
Netherlands	71.40	74.80
New Zealand	68.20	73.00
Norway	71.11	74.70
Spain	58.76	63.50

In table 2 below I have rounded all entries to the nearest whole year, reordered the table on the basis of longevity, and added a couple of spaces where they were needed. The result is an evocative picture of how different life is in developing nations. The gap between the life expectancies in Argentina and Brazil is now obvious and suggests that further investigation might reveal the causes for the difference.

TABLE 2
Life Expectancy at Birth

Country	Male	Female
Netherlands	71	75
Norway	71	75
New Zealand	68	73
Canada	68	73
Iceland	66	70
Japan	65	70
Spain	59	64
Argentina	57	61
Brazil	39	46
Mexico	38	40

Just as increasing the number of decimal places can make a table harder to understand, so can increasing the number of dimensions make a graph more confusing. No one interested in unambiguous data display would choose the form shown in figure 40 on which to build a graph.

We have already seen (rule 4) how extra dimensions can cause ambiguity (is it length or area or volume?). In addition, human perception of areas is inconsistent. Just what is and what is not confusing in a graph is sometimes only a conjecture, yet if the person who prepared the graph is confused, the viewer is almost certain to be. Shown in figure 41 is a plot of per share earnings and dividends over a six-year period. The person constructing the graph chose to make the bars three-dimensional (perhaps to keep them upright against the winds of economic change). We note that 1975 starts out OK but ends as the side of a bar. The third dimension of this bar chart has confused the artist!! A hint that something has gone amiss shows itself because a bar is left over after all the years have run out. I suspect that 1976 is really 1975, and the unlabeled bar at the end is probably 1977. A simple line chart with this interpretation is shown in figure 42.

FIGURE 40. A three-tined, two-pronged Blivet (from a cocktail napkin).

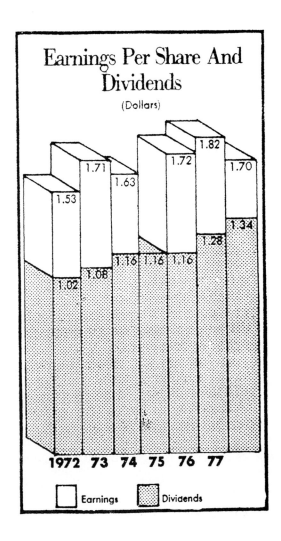

FIGURE 41. An extra dimension on earnings and dividends confuses even the grapher (from the *Washington Post*, 1979).

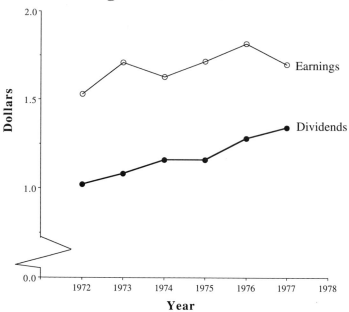

Earnings and Dividends Per Share

FIGURE 42. Data from figure 41 redrawn simply.

RULE 12   If it has been done well in the past, think of a new way to do it.

We have all seen maps that are shaded to represent some nongeographic variable; it might be cancer rates, or per capita income, or average number of years of education. Such maps have a long history. John Fletcher, an early developer of such maps, produced some in 1834 that remain favorites. Figure 43 uses shading to show the distribution of ignorance in England and Wales. I have long hoped for someone with suitable data to prepare a set of parallel plots for the United States over the last few decades.

Showing geographic distribution of a single variable with shading is not without its pitfalls,* but developing an effective display is pretty straightforward. To show median family income in Fort Worth, Texas, (see figure 44) the U.S. Bureau of the Census used variations in the saturation of red shading, using yellow to represent the lowest level. Of course one needn't use red; when mapping "percent of high school graduates" for the same region, they used saturations of blue to equally good effect (see figure 45). But suppose the goal is to show both variables simultaneously on a geographic background? One solution immediately suggests itself (especially if you are the Bureau of the Census and proud possessor of very snazzy software and plotting equipment): the two-variable color map. This map format (see figure 46) is essen-

*One of these, the so-called patch-map problem, is that the visual impact of a specific geographic region may be far out of proportion. For example, a slight elevation in, say, cancer rates, in a large, though sparsely populated area (like Nevada) suggests a more serious problem (in terms of number of people affected) than a much smaller increase in a small, densely populated state (New Jersey), even though the number of new cases might be many times more in the smaller region.

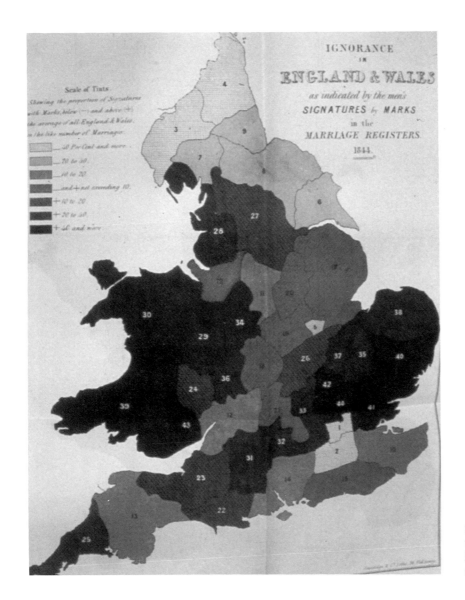

FIGURE 43. Fletcher's plot of the distribution of ignorance in nineteenth-century England and Wales.

tially a plotting of the two single-component maps on top of one another, letting the plotting colors mix. The four variations of each color translates to the four-by-four matrix of color mixtures shown in the legend.[8]

What is visible? The resulting map is undoubtedly beautiful, but for most of the kinds of questions we ordinarily would want to ask of such data (e.g., Is there any relationship between education and income? If so, how large?) it offers little help. Moreover, the two-variable version has effectively obscured what was clear in the one-dimensional components. (Also, after more than two hundred practice exercises with such maps, graduate students in perception at Johns Hopkins University were unable to internalize the legend.)

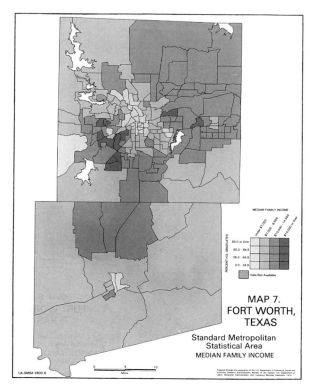

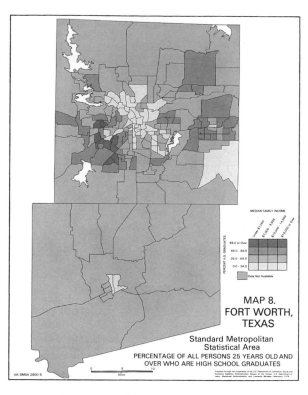

FIGURE 44. The geographic distribution of median family income in Fort Worth, Texas, in 1974. (See insert for color version.)

FIGURE 45. The geographic distribution of percentage of high-school graduates in Fort Worth, Texas, in 1974. (See insert for color version.)

FIGURE 46. The geographic distribution of both median family income and percentage of high-school graduates in Fort Worth, Texas, in 1974, shown as a two-variable color map. (See insert for color version.)

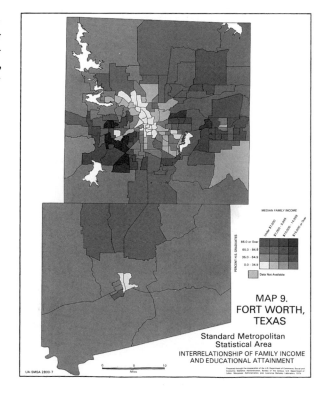

Not solving a four-dimensional display is nothing to be ashamed of. Census should be credited with a valiant beginning. It is ironic, however, to note that one hundred years before the census version a two-variable color map was done rather well by Georg von Mayr.[8] He used bars of different widths and frequencies (see figure 47) to depict the distribution of horses and cattle in eastern Bavaria, and in so doing accomplished gracefully what Census, using varying saturations, did clumsily. Census's two-variable color map is a wonderful example of how color in a graph can seduce us into thinking that we are communicating more than we are.

A particularly enlightening experience is to look carefully through the six books of graphs that William Playfair published between 1786 and 1822. One discovers clear, accurate, and data-laden graphs utilizing many useful ideas that are too rarely applied today.

Summing Up

The twelve "rules" for achieving graphical failure are only the beginning. Nevertheless, they point clearly toward an outlook that tells us much about good display. The measures of display interlock: the data

FIGURE 47. A two-variable color map showing the joint distribution of horses *(Pferde)* and cattle *(Rindvich)* in eastern Bavaria in 1874. (See insert for color version.)

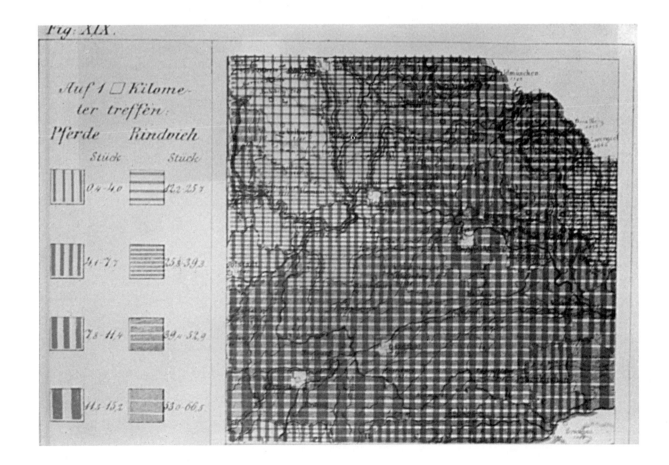

density cannot be high if the graph is cluttered with chartjunk; the data/ink ratio grows with the amount of data displayed; the lie factor manifests itself most frequently when additional dimensions or worthless metaphors are included.

Thus, the rules for good display are simple:

1. Examine the data carefully enough to know what they have to say, and then let them say it with a minimum of adornment.
2. In depicting scale, follow practices of "reasonable regularity."
3. Label clearly and fully.

Last, and perhaps most important, spend some time looking at the work of the masters of the craft. An hour spent with Playfair or Charles Joseph Minard will not only profit your graphical expertise but will also give you great pleasure. For technical details, Jacques Bertin's *Semiology of Graphics*[10] is without peer, and for an unabashed celebration of the possibilities of graphic display, beautifully produced, nothing touches Edward Tufte's three interlocking books.[11]

CHAPTER 2 Graphical Mysteries

"The greatest value of a picture," observed the renowned polymath John Tukey, "is when it *forces* us to notice what we never expected to see."

The obvious truth of this famous epigram has been discovered by all who explore data. Removing graphs from a data analyst's arsenal would make the task of discovery insuperably more difficult.

Sandy Zabell dramatically illustrated this truth in his reanalysis of the London Bills of Mortality,* which had been previously examined carefully by the seventeenth- and eighteenth-century scholars Arbuthnot, Brakenridge, and Graunt. Using graphical methods, Zabell found inconsistencies, clerical errors, and a remarkable amount of other information in the bills, "much of it unappreciated at the time of their publication" (p. 27). Indeed, when Zabell graphed these data as a time series, the errors stuck out, literally, like sore thumbs.

How were they missed by these earlier researchers? The solution to this first graphical mystery is easy; graphs were not widely known until the publication of William Playfair's influential *Commercial and Political Atlas* in 1786.

Before Playfair, the use of data graphics, although not completely unheard of, was rare. During the eighteenth century no graphs were to be found in any volume of any of the following journals:

Acta Eruditorum

Annals of Philosophy

Edinburgh Journal of Science

Edinburgh Philosophical Journal

Mémoires de l'Academie des Sciences (Paris)

Mémoires Présentés par Divers Savans à l'Académie Royale des Sciences (Paris)

Journal of Natural Philosophy

*In London of the 1530s parish clerks were required to submit weekly reports on the number of plague deaths. These bills of mortality were meant to tell authorities when measures should be taken against the epidemic. The publications of the *London Bills of Mortality*, a summary of these individual reports, was begun by the Company of Parish Clerks in 1604.

Novi Commentarii

Academiae Scientiarium Imperialis Petropolitanae

Observations sur la Physique

Philosophical Transactions of the Royal Society of Edinburgh

The only European journal containing graphs during this entire century was the *Mémoires de l'Academie Royale des Sciences et Belle-Lettres* (Berlin), in papers by Lambert in the mid- to late eighteenth century (about thirty line graphs showing a variety of physical phenomena like evaporation rates) and one paper by Benjamin Thompson[1] in an article on ballistics. Joseph Priestley used a bar graph in 1765 to depict the life spans of some two thousand celebrated persons who lived from 1200 B.C. to 1750 A.D. Interestingly, Priestley felt it necessary to write several pages of explanation to justify—as a natural and reasonable procedure—his representation of time by a line in his charts. (Subsequently, he must have found the justification unnecessary, since he omitted it in his 1769 elaboration.) His graphs were the only ones Playfair acknowledged as a precedent.

The power of human beings to decode information when it is presented graphically is so enormous that we are shocked when it fails. We ought not to be, since it is analogous to the Peter Principle in that as graphs are drawn worse and worse, their messages remain visible, until finally they fail and the message it carries is lost. But because other dreadful graphs were understood in the past, the perpetrator doesn't consider the possibility of failure and assumes instead that there is no message. This is the only explanation that makes sense to me for the proliferation of computer software that produces such nonsense as pseudo-three-dimensional pie charts in 256 colors.

Following are three graphical failures. Not only do they fail to force their message upon us, but even after we know what to look for, they still do not allow the message to be seen. In the first two, the graphical sins were venial in that the consequences were only that the graphs, in Alfred North Whitehead's memorable phrase, "left the vast darkness of their topic unobscured." The consequences devolving from the third were tragic.*

Mystery #1: Pregnancies Among !Kung Women†

!Kung women in Botswana tend to go about three years between pregnancies. They do this without birth control. How? Konner and Worthman[2] constructed figure 1 to support their theory that frequent nursing maintains a high level of progesterone, a natural hormone that prevents ovulation. Does it convince you?

*Will Rogers's comment, "What we don't know won't hurt us, it's what we do know that ain't," was never more clearly appropriate.

†Examples 1 and 2 were taken from Bill Cleveland's important and useful book *The Elements of Graphing Data*. Example 3 comes from Dalal, Fowlkes, and Hoadley's scholarly paper (1989), whose implications were brought home to me by Edward Tufte during a workshop he gave more than a year ago. There is an elegant description of the Challenger disaster in his new book *Visual Explanations*. I am grateful to Bill and Edward for their permission to use the results of their labor.

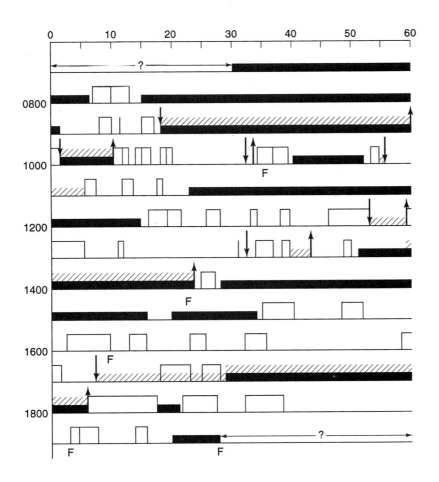

FIGURE 1. The graph shows the activities of a !Kung woman and her baby. The open bars and tall vertical lines are nursing times; the closed bars show times when the baby is sleeping; F means fretting; and slashed lines are intervals when the baby is held by the mother, with arrows for picking up and setting down. (Redrawn from Cleveland, 1994, p. 25. Reprinted with permission.)

Konner and Worthman's graph is confusing because

1. It uses different symbols to represent different events on a common time scale. Irrelevant events have at least as much visual weight as the theoretically important ones.
2. Understanding the graph requires memorizing a complex legend and ignoring everything superfluous.

Redoing the graph, using a common visual metaphor for the occurrence of an event, relieves the viewer of the onerous task of learning the legend and one can focus attention on the variable of interest, nursing.

It is known that nursing increases the body's production of progesterone, but the effect is very short-lived, and so a woman must nurse almost hourly to keep the progesterone level high enough for it to be an effective contraceptive. A quick glance at the top row of each panel in figure 2 shows that indeed !Kung women seem to be nursing frequently enough to maintain high progesterone levels. But even if there were no link between nursing and progesterone production, the nursing frequency seems to leave little spare time for the activities required to become pregnant.

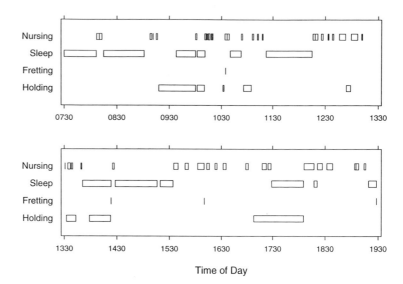

FIGURE 2. The data from figure 1 are regraphed using a common metaphor for all activities. It is now easier to see the activity times and their interactions. (Reprinted, with permission, from Cleveland, 1994, p. 27.)

Mystery #2: The Three Mile Island Accident

At the end of March 1979 an accident occurred at the Three Mile Island nuclear reactor that resulted in the release of some contaminants into the atmosphere. One of these contaminants was isotope 133 of the gas xenon. Samples of air were gathered at about the same time in Albany, 375 kilometers downwind from Three Mile Island. Do these samples show elevated readings of xenon 133? If so, how elevated? If elevated, how long did the contaminants take to travel the 375 kilometers?

In a 1980 article in the journal *Science,* Wahlen, Kunz, Matuszek, Mahoney, and Thompson tried to communicate the answer to these questions with the graph shown as figure 3. This graph is difficult to understand because of a multiplicity of scales, unusual and unexplained conventions, and a confusing mixture of data types within the same plot.

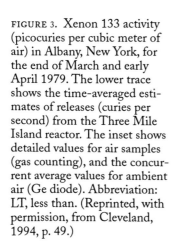

FIGURE 3. Xenon 133 activity (picocuries per cubic meter of air) in Albany, New York, for the end of March and early April 1979. The lower trace shows the time-averaged estimates of releases (curies per second) from the Three Mile Island reactor. The inset shows detailed values for air samples (gas counting), and the concurrent average values for ambient air (Ge diode). Abbreviation: LT, less than. (Reprinted, with permission, from Cleveland, 1994, p. 49.)

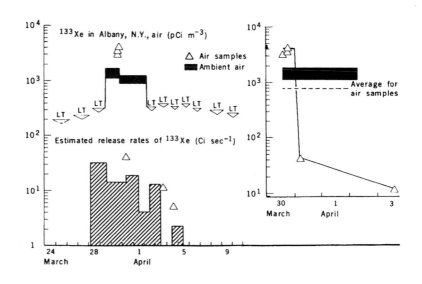

<blockquote>
SECTION II
</blockquote>

Graphical Triumphs

So far we have seen a catalog of graphical disasters—displays that have left the viewer as uninformed after viewing them as before. If this is all that we have to show from our graphical heritage it would be a waste of time to write or read a book about such a worthless tool. While graphical flops are too common, the triumphs of graphical displays are so remarkable and so important that it makes their misuse tragic. Images of a snowplow on the front of a Ferrari, Secretariat hitched to a Central Park wagon, and Mozart deciding on accounting as a profession all are appropriate metaphors for much of contemporary graphical use.

Good graphs can make difficult problems trivial. We have all become used to weather forecasts that are very accurate and detailed for a day or two and pretty good for as far in advance as a week. I used to think that this was due to the increasing sophistication of complex prediction models.* But then I noticed the weather maps shown on every news broadcast. Using a model of no greater sophistication than that employed by Benjamin Franklin (weather generally moves from west to east) I was able to predict that the area of precipitation currently over Ohio would be hitting New Jersey by tomorrow and would stay over us until the weekend. Any fool could see it. The improvement in forecasting has not been due to the weather models. The enormous wealth of radar and satellite data summarized into a multicolored and dynamic graphic can turn anyone into an expert. The horrors chronicled in section I are the result of the misuse of this marvelous tool.

*To some extent it is certainly true that models are more sophisticated. I was enormously impressed when some surprising turns in a hurricane's path were predicted well in advance, but such models seem to be needed no more often than seldom.

In the next five chapters some of the most wonderful graphics in history will be reproduced and described. In looking at them go slowly. Care will be rewarded, for their beauty is more than skin deep.

In chapter 3 we begin with a reminder of how an old map foretold a modern theory, we pass through some graphical evidence that settled a scientific disagreement, examine a marvelous graphical solution to a vexing and subtle problem of airplane design, and end with John Snow's map of a nineteenth-century cholera epidemic.

Chapter 4 contains what may be the best graph ever drawn—Charles Joseph Minard's plot showing Napoleon's march into Russia. A careful examination of this graph and a minute's thought about what is being conveyed will bring a lump to the throat of even the most hardened. Paired with Minard's graph is the deceptively simple line graph that Norman Maclean used to communicate both the terrible tragedy at Mann Gulch and a path toward avoiding such unnecessary sacrifices of young lives in the future.

Chapter 5 contains a couple a common graphical procedures in physics, but culminates with Richard Feynman's graphical solution to the problems of quantum electrodynamics (QED). His graphical method transformed a set of mathematical procedures that were so formidable that almost no one could use them (Julian Schwinger, who shared the Nobel Prize with Feynman, being one notable exception) into a tool taught in graduate school to all students interested in navigating the strange world of quantum theory. The Feynman diagram did for QED what dynamic maps and augmented satellite photos have done for weather forecasting.

Chapters 6 and 7 both deal with the seemingly mundane matter of train and bus schedules. The central idea is that through the use of a graphical rather than tabular format the reader can quickly gain a complete understanding of the structure of the trains and buses. They are also easier to read without glasses, less likely to be destroyed through repeated photographic reproductions, and more likely to get you to your destination, when you want to get there.

The graphs in this section are a celebration of some of the very best information displays ever invented. They inform us about the structure of the universe, get us where we want to go, and memorialize lost lives; they show how a statistical graphic, when carrying important information and wisely constructed, can be so evocative as to be almost poetic.

CHAPTER 3 Graphical Answers to Scientific Questions

Although there have been many contributors to the development of graphical methods for the depiction of data, William Playfair (1759–1823) was the most influential of innovators. He was a popularizer and propagandist whose inventions found immediate acceptance because they worked so well. In his own somewhat immodest words,

> I found the first rough draft gave me a better comprehension of the subject, than all that I had learnt from occasional reading, for half my lifetime.
>
> —Advertisement on prelims, *An Inquiry*, 1805

The unrelenting forcefulness inherent in the character of a good graphical presentation is its greatest virtue. We can be forced to discover things from a graph without knowing in advance what we were looking for.

There are many examples of important discoveries in which graphics have played a vital role. In selecting four to present here, I chose a strategy (that may appear to be overkill) to counteract the common misunderstanding of the role of graphs in the development of scientific theories.

Example 1: Continental Drift

In a history of experimental graphs, Margaret Tilling restates this misconception:

> Clearly an ability to plot an experimental graph necessarily precedes an ability to analyze it. However, although any map may be considered as a graph, and carefully constructed maps had been in use long before the eighteenth century, we do not expect the shape of a coastline to follow a mathematical law. Further, although there are a great many physical phenomena that we do expect to follow mathematical laws, they are in general so complex in nature that direct plotting will reveal little about the nature of those laws . . . .[1]

Attitudes like these have hindered appropriately serious regard for such theories as that of continental drift, my first example, whose initial evidence (noticed by every schoolchild) is solely graphical (see figure 1).

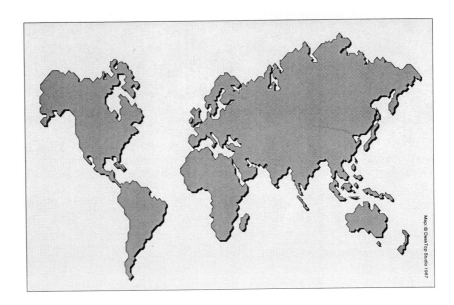

FIGURE 1. A familiar map projection that fairly screams "continental drift."

Example 2: Did Neanderthals Talk Like Pigs?

Paleontologists have argued for more than a century over whether Neanderthal man could talk. This issue became especially important recently in the debate about whether the Neanderthals were an integral part of the human developmental line or an evolutionary dead end. In 1989, a description of the only known Neanderthal throat bone, the hyoid, was published by Baruch Arensburg and his colleagues at Tel Aviv University. This 60,000-year-old bone was found in Kebara cave, near Mt. Carmel, in Israel. Arensburg provided six measurements of the bone and concluded that they compared favorably to similar measures of the hyoid of modern humans. This conclusion was disputed by Jeffrey Laitman and Joy Reidenberg, who showed that pig hyoids are remarkably similar to the Kebara hyoid on two of the six measurements.[2]

This argument was apparently settled when David Frayer, of the University of Kansas, showed a photograph of human, Neanderthal, and pig hyoids at a recent meeting in Toronto.[3] See figure 2.

Example 3: Armoring Airplanes

Abraham Wald, in some work he did during World War II that only became available thirty-five years after the end of the war,[4] was trying to determine where to add extra armor to planes on the basis of the

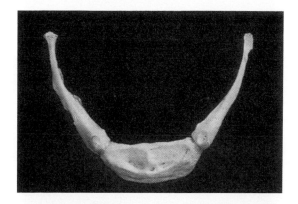

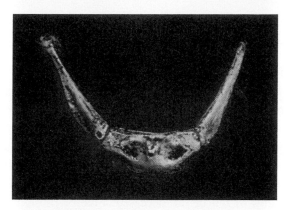

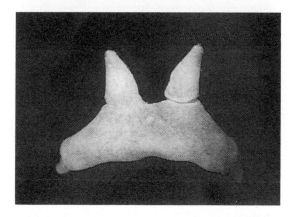

FIGURE 2. Hyoid bones from a modern human (*top*), Neanderthal (*middle*), and pig (*bottom*).

pattern of bullet holes in returning aircraft. His conclusion was to determine carefully where returning planes had been shot and *put extra armor every place else!*

Wald made his discovery by drawing an outline of a plane (crudely shown in figure 3) and then putting a mark on it where a returning aircraft had been shot. Soon the entire plane had been covered with marks *except* for a few key areas. He concluded that since planes had probably been hit more or less uniformly, those aircraft hit in the unmarked places had been unable to return, and thus were the areas that required more armor.

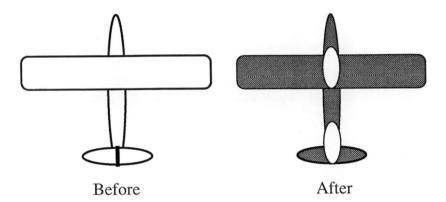

Before   After

FIGURE 3. Abraham Wald drew his incredible conclusion about armoring airplanes only after he drew "maps" of bullet holes on returning aircraft.

Example 4: The Source of a Cholera Epidemic

Dr. John Snow plotted the locations of deaths from cholera in central London in September 1854 (figure 4). Deaths were marked by dots, and in addition, the area's eleven water pumps were located by crosses. Snow observed that nearly all of the cholera deaths were among those who lived near the Broad Street pump. But before he could be sure that he had discovered a possible causal connection, he had to understand the deaths that had occurred nearer some other pump. He visited the families of ten of the deceased. Five of these, because they preferred its taste, regularly sent for water from the Broad Street pump. Three others were children who attended a school near the Broad Street pump. On September 7 Snow described his findings before the vestry of St. James Parish. The graphic evidence was sufficiently convincing for them to allow him to have the handle of the contaminated pump removed. Within days the neighborhood epidemic that had taken more than five hundred lives had ended.*

At the time Snow did his investigation, very little was known about the vectors of contagion of disease. Theories of "foul vapors" and "divine retribution" were still considered viable. The map that resulted from Snow's methodical work did not uncover the bacterium *Vibrio cholerae,* which current theories consider cholera's cause, but it drew the causal connection between the transmission of cholera and drinking from the Broad Street pump. His work is often cited as an early example of what has grown into modern epidemiology.

*The Broad Street pump is now gone. In its place is the John Snow Pub. See Gilbert (1958) and Jaret (1991) for more details.

Graphs Have Not Always Been Around

Graphs are so basic to our understanding that we cannot easily imagine the world without them. This was brought home to me some years ago when I was reading a technical report by Zabell[5] that examined the *London Bills of Mortality* and their analysis by three early statisticians.[6] The aim of the paper was

to see how much these writers were able to extract from the Bills that we might reasonably expect them to—for example, how sensitive they were to questions of data quality, data consistency and data aggregation—we deliberately avoid the use of modern statistical methods . . . and limit ourselves to what is, in effect, a simple form of data analysis.

The result of these simple analyses was that a variety of errors were discovered that should have been seen by these early investigators, but were not. Zabell concluded:

Although we have deliberately avoided all but the simplest of statistical tools, a remarkable amount of information can be extracted from the Bills of Mortality, much of it unappreciated at the time of their publication.

FIGURE 4. A map constructed by John Snow in September 1854 showed that most of the deaths due to cholera clustered around the Broad Street water pump. It is often used as a landmark in epidemiology. (From Tufte, 1983. Reprinted by permission.)

The "simple" methods of data analysis he used were graphical. Such data characteristics as clerical errors in the *Bills* literally stuck out like sore thumbs. Yet Zabell's carefully researched work was flawed. The graphical method on which his analysis leans so heavily was developed after the scholars he discussed did their work. Thus despite his desire to play eighteenth-century scholar and use only techniques of analysis available at the time, Zabell fell into an anachronism. This incorrect assumption is but one indication of how ubiquitous the notion of graphical depiction has become; it is hard to imagine the world without it.

CHAPTER 4 Three Graphic Memorials

"Hear, forget; see, remember." The wisdom of this ancient Confucian saying is apparent. Memorable memorials are visual. Who can ever forget the tragedy chronicled by the austere black granite wall that is the Vietnam Memorial? It is massive in form and content, built from the space taken by the more than 58,000 names inscribed upon it. As the loss of life increases, so too does the height of the wall, and the emotions it evokes. It is a very personal thing. William A. Atwell, Terry Lee Dillard, Ward K. Patton, Jerry Lee Graves, Edward J. Downs, John E. Rice, Jack M. Strong—these names join with thousands of others to form the wall. The interaction of the monument with those who come to it, whether to seek out a particular name or to picnic, often becomes part of the diverse images we take away with us. The tragedy of Vietnam written in the small becomes large and indelible.

Napoleon's Russian Campaign

Memorializing that portion of the generation of young French men lost in Napoleon's ill-fated Russian campaign was surely part of Charles Joseph Minard's motivation in the construction of his famous 1869 graphic. Minard's plot, shown in figure 1, depicts the movement of the French army from the time it crossed the Polish-Russian border with 422,000 men in June of 1812. The shrinking size of the army is characterized by the progressive narrowing of the broad band stretching across the map. In the original scale, each millimeter of its width represents 6,000 men. When the army reached Moscow in September, only 100,000 remained. The city was deserted, and the army began its retreat, depicted by the darker line below. It is linked to the temperature scale showing quantitatively the depths of the Russian winter. The banks of the Berezina River were littered with the bodies of the 22,000 men who perished as the November temperature

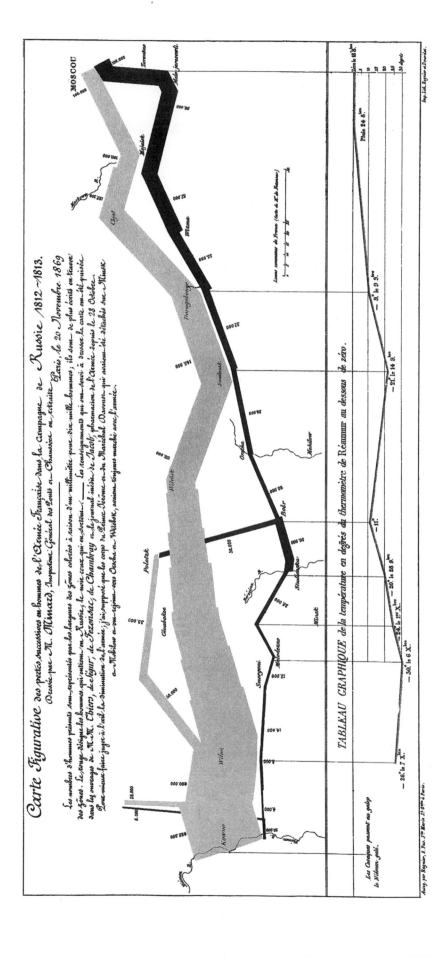

FIGURE 1. Charles Joseph
Minard's famous plot of
Napoleon's Russian Campaign.
(Reprinted, with permission,
from Tufte, 1983.) (See insert
for color version.)

dropped to −20°. When the remainder of the army crossed into Poland as the year ended, only 10,000 men remained.

The story of the tragedy is clear. We can see the bodies frozen into the snow. Marey told how this graph "brought tears to the eyes of all France."[1] No wonder; there were few families unaffected.

Minard's depiction of Napoleon's Russian campaign has been characterized as perhaps "the best statistical graphic ever drawn."[2] Why? It is not the quality of the pen stroke, although it certainly passes muster in that regard. It is the importance and richness of the data. A single page carries six variables that tell the evocative story of where and how thousands of men died. Its poignancy is heightened through the immediate and graphic answer to the question, Compared to what? Ten thousand men returned. A lot or a few? Opposing the returning trickle against the departing torrent answers the question. The difference between them measures the tragedy. But nowhere does the shrinking distance between two lines depict a more touching tragedy than in my next example.

The Mann Gulch Fire

A smaller, and until recently lesser-known, tragedy took place on August 5, 1949, in the Helena National Forest, in Montana. August is amid the worst part of the dry season, so when lightning struck the trees near the Meriwether Guard Station, it wasn't surprising that a fire broke out.

The fire spread quickly north to that portion of Mann Gulch that abuts the Missouri River. Once the fire was spotted, sixteen Forest Service smoke jumpers were parachuted in to fight the blaze. The area around Mann Gulch is very rough country, and when the smoke jumpers were able to get close enough to judge the fire (point 6 on the map shown as figure 2), they realized that it was beyond their control and that their survival depended on quick action.

The fire was moving rapidly upgulch (northeast), so the smoke jumpers headed upslope, perpendicular to the fire's motion. They ran furiously, trying to get over the ridge that demarcated Mann Gulch ahead of the fire that was burning at their backs. Figure 2 is an annotated topographic map of the Mann Gulch area. The contour lines indicate that the slope was approximately seventy-eight percent in the region near the ridge at the top of Mann Gulch. The elevation of 4,700 feet marks the boundary of safety. The bodies of thirteen of the young men were found on this steep escarpment—within two hundred yards of the haven at the top of Mann Gulch.

The story of the Mann Gulch fire was told by Norman Maclean, who gathered information from many sources to reconstruct the event.[3] He augmented the already horrifying map with a graph whose

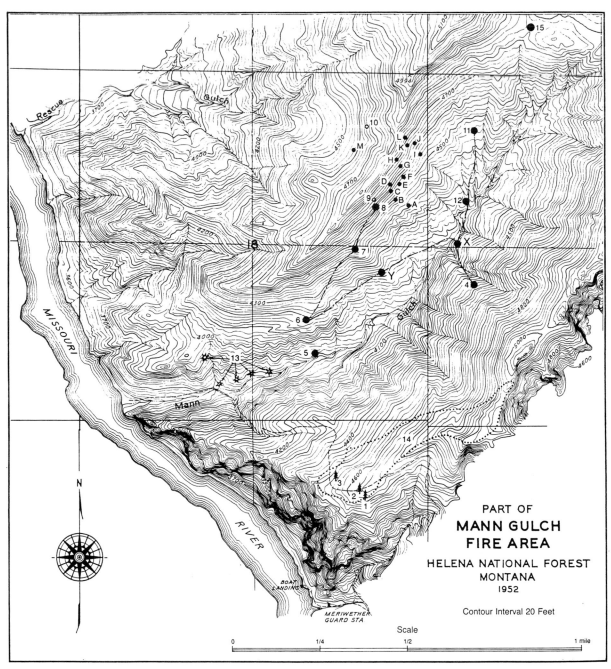

PART OF
**MANN GULCH
FIRE AREA**

HELENA NATIONAL FOREST
MONTANA
1952

Contour Interval 20 Feet

Scale

0   1/4   1/2   1 mile

FIGURE 2. A map of part of
the Mann Gulch Fire area
augmented with information
about the smoke jumpers who
came to fight the fire. This was
reprinted from Maclean (1992),
which in turn reprinted it from
a 1952 U.S. Department of
Agriculture report.

LEGEND

1,2,3	Lightning struck trees.
4	Dodge met Harrison.
X	Dodge ordered crew to north side of Gulch.
Y	Dodge and Harrison rejoined crews; beginning of crew's race.
5	Jansson turned back.
6	Dodge and crew turned back.
7	Dodge ordered heavy tools dropped.
8	Dodge set escape fire.
9	Dodge survived here.
10	Rumsey and Sallee survived here.
11	Jumping area (chutes assembled, burned).
12	Cargo assembly spot (burned).
13	Spot fires.
14	Approximate fire perimeter at time of jumping and cargo dropping (3:10–4:10 P.M.).
15	Helecopter landing spot.

BODIES FOUND

A	Stanley J. Reba
B	Silas R. Thompson
C	Joseph B. Sylvia
D	James O. Harrison
E	Robert J. Bennett
F	Newton T. Thompson
G	Leonard L. Piper
H	Eldon E. Diettert
I	Marvin L. Sherman
J	David R. Navon
K	Phillip R. McVey
L	Henry J. Thol, Jr.
M	William J. Hellman

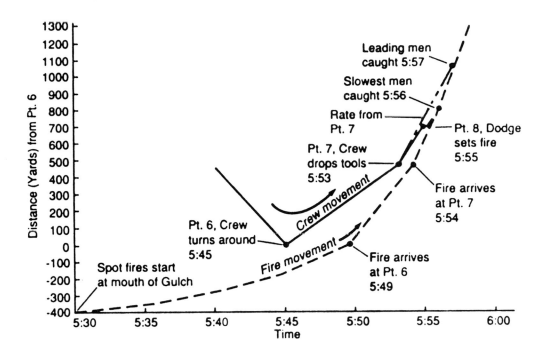

austerity contrasts with the tragedy it conveys (figure 3). In Maclean's words,

> The tragic convergence of fire and men in Mann Gulch offers itself as a tragic model for a graph, the modern scientist's favorite means of depicting what he wishes to present as clearly as possible. Drawn along axes of time and distance is one line depicting the course of the fire and one depicting the course of the men, and where there is convergence of the two, graphically speaking, is the tragic conclusion of the Mann Gulch story; the two lines converging to this conclusion constitute the plot.

FIGURE 3. Time and position of crew at Mann Gulch on August 5, 1949. (Graph by Richard C. Rothermel. Taken from page 269 of Maclean, 1992. Reprinted with permission of the University of Chicago Press.)

The rapid upturning of the dashed line conforms to the characteristic spread rate of fire, increasing in proportion to the square of the slope of the terrain—a stark contrast to the effect of steep slope on the swiftness of human ascent. It was very steep where the young men died. Francis Bacon's apothegm *"Old men go to death, but death comes to young men"* was never more apparent.

As we study the graph, the character of the race between men and fire becomes more vivid still.

> Along each line are numbers which are turning points in the race between men and fire, and if the lines are viewed as a race the numbers mark off legs of the race, if they also have religious significance they are stations of the cross, and if they have literary significance they mark off acts of drama, . . . but the acts are short, because modern wildfire allows no time for soliloquies.[4]

The impact of a memorial is built from many parts. Largest, of course, is the human tragedy chronicled. But what makes for a tragedy? The dictionary says that it is a disastrous event, but a richer definition must include a losing battle against inexorable forces. This last graph makes clear the unrelenting quadratic acceleration of the Mann Gulch fire. It also fosters painful thought experiments. Safety is represented by the undrawn 1400-yard line. If only the crew's line could be translated back five minutes or up 200 yards . . . .

Despite their memorial similarities, these three graphics are different in their goals. The Vietnam Memorial's message is largely emotional (although certainly with practical consequences), emphasizing the costs of war. Minard's plot does this as well, but it also conveys a considerable amount of information about the many dimensions contributing to the failure of Napoleon's campaign. The Mann Gulch plot does something special. By making explicit the quantitative connection between the slope of the land and the speed with which fire spreads upon it, it allows wiser decisions about where and when to fight forest fires. It uses the tragedy of Mann Gulch directly to help prevent future deaths.

"See, remember."

CHAPTER 5 A Nobel Graph

Introduction

The map is the earliest schematic spatial representation to evolve; the geometric diagram, being more abstract, appeared later. Cartography and geometry were historically the most important areas of graphic development. Geometry evolved to treat space as a pure abstraction equivalent to number. The movement of objects in space* made for ready extension to include time in the graphics spatial plane. Other graphic systems, including musical notation and astrological and chemical symbol systems, were all partially diagrammatic in character. The systemization of the spatial diagrammatic forms by Descartes in 1637 into the Cartesian coordinate system was an integration of the geometric and algebraic systems, and established what remains the most intellectually important and useful of diagrammatic graphic systems.

The Cartesian system so dominated intellectual conceptions of what graphs were and what they were for—that is, the depiction of the mathematical functions governing the behavior of objects in space and time—that it took more than a century and a half before it occurred to anyone that graphs could be used for anything else.[1] Furthermore, the Cartesian tradition was so strong that it misled those who were using graphs in entirely different ways for altogether different intellectual tasks into the belief that they were doing Cartesian geometric analysis. In truth, they were engaged in something quite different, involving no

*An early example of a space-time graph of empirical data within the statistical tradition was an article on ballistics by Benjamin Thompson in 1782 in the *Philosophical Transactions*. Earlier, but more Cartesian in its goals, was Kepler's curve fitting of the astronomical data he inherited from Tycho Brahe. This 1609 work yielded the first two of what are now known as Kepler's Laws.

Portions of this chapter have previously appeared as an article written in collaboration with Professor John W. Durso of Mt. Holyoke College. I am grateful for his permission to reprint it here.

geometry more complex than that well known even in preclassical antiquity. What they were doing was, however, an important intellectual departure from Cartesianism—that is, graphic methods were being applied to exploratory analysis of empirical statistics. Eventually, with d'Alembert, Gauss, and others, a method of joining the Cartesian and the statistical graphic approaches developed, but in application, curve-fitting remained on an intellectually separate track until very late in the nineteenth century. Cartesian curve-fitting uses data to determine the structure of the laws governing the universe; the fitted curves are considered to be empirical manifestations of those laws. The statistical orientation uses curves (regularities) to determine the structure of concrete sets of data—data about phenomena that are important to understand in their own right.

The principal focus of this chapter is clearly in the Cartesian tradition. Physics has long been in the forefront of those sciences that utilize a spatial representation to solve problems. Finding the solution to even the most rudimentary of mechanics problems is eased through the use of a parallelogram to calculate the resultant of two forces. But there are other, more interesting examples.

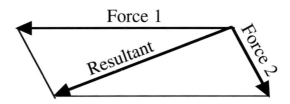

*Nomogram. From the Greek *nomos* = law. Literally, a diagram that conveys a law.

We begin this exposition of nomograms* with a more traditional example of how the geometry of the circle was used by Galileo to derive and extend known physical laws. We then jump forward three hundred years and describe how Richard Feynman became a Nobel laureate by inventing a graphical tool to extend the methodology surrounding the theory of quantum electrodynamics (QED). His nomogram allowed QED to emerge as a practical, working physical theory and provided insights into the nature of all fundamental interactions.

Galileo's Problem†—How Inclined the Plane?

†This example was suggested by Peter Cheng and Herbert Simon's chapter "Scientific discovery and creative reasoning with diagrams," Smith, Ward, and Finke (1995).

Consider (figure 1) trying to roll a ball down the inclined plane from the starting line to the goal in the shortest possible time. Galileo knew that the steeper the incline (the larger the angle θ), the faster the ball will roll. But the steeper the slope, the longer the distance the ball will have to travel. How can we determine the optimal angle? We clearly see that the distance is minimized when the angle is zero, but then the ball won't roll at all (time of travel will be infinite). Similarly, when the angle is 90° the speed is maximized but the length of the trip is infinite. Obviously, the solution lies between these two extremes. Both Galileo and Newton knew enough trigonometry to set up an equation that relates time of travel to angle of inclination. But only Newton could have taken its derivative, set it equal to zero, and solved it straightaway. Galileo didn't have calculus—only geometry.

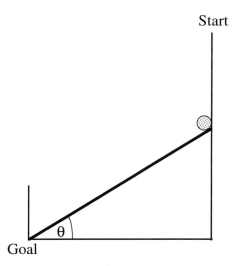

FIGURE 1. A graphical statement of Galileo's problem.

Galileo knew* that the time to traverse an inclined plane when it is inscribed in a circle is the same regardless of where on the circle one begins (see figure 2). Thus the fast fall but long distance from point D takes the same amount of time as the short slow trip from point A.

*Through a largely geometric argument essentially identical to the one explaining why the period of a pendulum depends only on its length and not on how far back it is pulled.

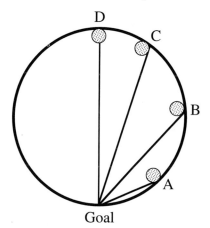

FIGURE 2. A graphical fact about rolling balls and circles.

Solving the original problem requires only combining figures 1 and 2 (into figure 3) and noting that all inclined planes but one involve some travel outside the circle (denoted by dotted lines). The point of tangency ($\theta = 45°$) must yield the minimal travel time. Of course we must go further to answer the obvious next question, How long does it take? But again Galileo's nomogram guides us to the solution. Since we know that the transit time will be the same from any point on the circle, we can easily calculate the time from the vertical position D in figure 2.

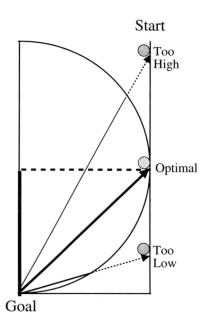

FIGURE 3. A graphical solution to Galileo's problem.

Feynman Diagrams—A Graphical Entrée into the Strange World of Quantum Electrodynamics

The quantitative stage of solution to a problem in physics is almost always writing down a set of equations that try to describe the system being studied and then solving them. Pictorial devices, such as the parallelogram of forces, are often critically needed guides over the usually complex route to a quantitative solution. For example, to add two forces analytically so that the sum can be expressed numerically, one would have to refer the forces, represented by the arrows, to some set of orthogonal axes, use trigonometry to resolve the forces into orthogonal components, add the corresponding components, and then again use trigonometry to reassemble the components showing both the correct magnitude and direction. The parallelogram drawn serves not as a direct calculative tool, but as an important guide to how to do the calculations.

The rapid development of quantum mechanics as *the* theory of matter in the decades following World War I presented physicists with the equations, but not the pictures. Despite this lack, physicists set out with vigor and diligence to work out the quantum theory of the interaction of charged particles with the electromagnetic field and with one another. By 1930, the fundamental equations of quantum electrodynamics had been worked out. But actually doing the calculations remained a task of considerable difficulty.

In 1949, the *Physical Review* published sequentially a pair of brilliant papers by Richard Feynman in which he reformulated quantum electrodynamics in a way that provided a deep, intuitive insight into the problem. To this day Feynman's work provides the most powerful guide to the calculations and the most useful view not only of electrodynamics but of all fundamental interactions.

Feynman's basic approach was to focus directly on the primary problem of the calculation of the probability* for particles in some specific initial state to interact and end in another specific state. As a consequence of his approach, which diminished the constraints imposed by a unidirectional concept of time, all particle dynamics in QED could be reduced to three fundamental actions. In addition, some side issues that are not experimentally accessible disappeared.

His result for the interaction of two electrons, for example, is a relativistic generalization of the familiar electrostatic repulsion of two like charges, modified to take into account the spin of the electrons. The general result was a simplified prescription for calculating the probability for any process involving electrons, positrons, and photons. The probability is expressed as a power series where each term in the series is evaluated by drawing a set of diagrams according to prescribed rules and associating specific mathematical factors with different features of the diagrams.

The mathematical factors are moderately complicated, but the rules derive from three simple actions that take place in electrodynamics:[2]

Action #1: A photon goes from place to place.

Action #2: An electron goes from place to place.

Action #3: An electron emits or absorbs a photon.

The rules assign a mathematical quantity to each of these actions. The probability of a specific process involving electrons, positrons, and photons is simply a sum of all the ways that the three possible actions can be arranged to get from the initial state to the final state of the process.

Suppose we wish to know what happens when a photon—a quantum of light—encounters an electron. This process is known as the Compton effect, and its successful description was crucial in the development of quantum mechanics. A critical step toward this ultimate goal is the calculation of a number that represents the probability of a system of particles moving from some initial state to some final state. The Feynman diagrams in figure 4 show the simplest ways for this to happen.

A solid line with an arrow represents an electron; a wavy line represents a photon. The sense of time flowing upwards is not to be taken completely literally, but only in the sense that the outgoing electron

*Actually he calculated the *transition amplitude* whose absolute square is the probability for the process to occur.

and photon are measured at a much later time than the incoming ones. The Feynman diagram in the left panel of figure 4 describes the process of an electron meeting and absorbing a photon at point 1, then moving to point 2 and emitting a photon. The theory, QED, not only describes the process qualitatively, but also predicts precisely the characteristics (energy, momentum, and spin) of the photon and electron at the end of the process from their values at the beginning.

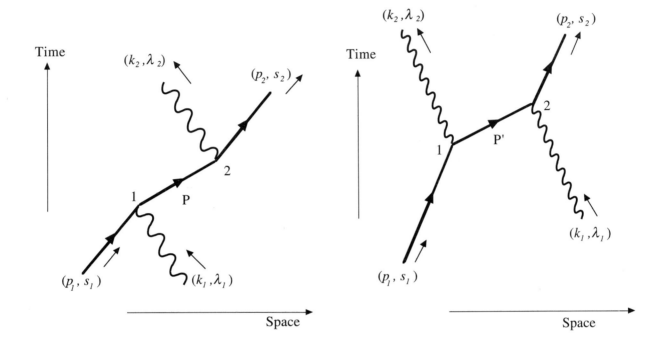

FIGURE 4. Lowest order Feynman diagrams for the Compton effect, an illustration of the graphical space-time approach to quantum electrodynamics. In the diagram, p_i and s_i are the 4-momentum and spin, respectively, of an electron at point i; k_i and λ_i are the 4-momentum and polarization of a photon at point i.

The drawing of this diagram requires that we understand the character of the phenomenon qualitatively. But after carefully constructing the diagram, we can transform our qualitative knowledge into a precise analytic description of the event by following the rules Feynman laid out. The details surrounding how one does this are somewhat arcane, but interested readers are referred to Durso and Wainer's careful description of the solution to this one simple example.[3]

Of course, the whole story of QED is more complicated than we have indicated, but the basic ideas expressed in the three possible actions remain at the heart of the theory. In 1965, Richard Feynman's contribution was recognized with the awarding of the Nobel Prize in physics (shared with Julian Schwinger and Sin-Itiro Tomonaga). While his view of electrodynamics was ". . . simply a restatement of conventional electrodynamics," his diagrammatic approach has become today's common language of quantum field theory. Nearly half a century after his insight, physicists are still thinking about fundamental interactions in the graphical way that Feynman taught.

CHAPTER 6 Todai Moto Kurashi*

I spent a day recently with Edward Tufte, whose three books on the display of information have served to set a new standard for elegance of design. The occasion was a one-day workshop on information display that Tufte gave in July 1995. In the course of this workshop I learned that with graphic displays, as with everything else, first impressions are sometimes deceiving.

The central display of this chapter is one that might initially appear off-putting, but with a little attention it becomes increasingly attractive: Marey's famous train schedule.

Tufte is haunted by timetables. Nineteen timetables (train, plane, and operatic) appear in his book *Envisioning Information.* They form such a steady theme that they prompted Bill Eddy, in his otherwise enthusiastic review of the book, to exclaim "But Ed, enough is enough." The number of people in the street who would characterize a train schedule as beautiful must surely be limited, but sensing true inner beauty always requires intimacy.

Most of the nineteen schedules and timetables that Tufte presents are variations on the theme that first appeared more than a hundred years ago when E. J. Marey published Ibry's now famous graphical train schedule (figure 1). In it the time of day is depicted along the horizontal axis and the stations are spaced proportionally along the vertical axis. Trains from Paris to Lyons are naturally shown as lines descending from the left to the right. Trains in the opposite direction ascend from left to right. The metaphor is complete; at a glance it shows both the general and the particular. There are thirty-eight trains and thirteen stations. Train departures from Paris are at their most frequent in the early evening. Steeper slopes denote faster trains. We can see that the twenty minutes of extra sleep obtained by preferring the 6:50 train from Paris over the one at 6:30 is paid for by increasing the length of the journey to Lyons by five hours.

Tufte used Ibry's scheme to build a graphical schedule to describe the more than two hundred buses connecting Hoboken and New

*The chapter title is an ancient Japanese saying: "It is always darkest under the light."

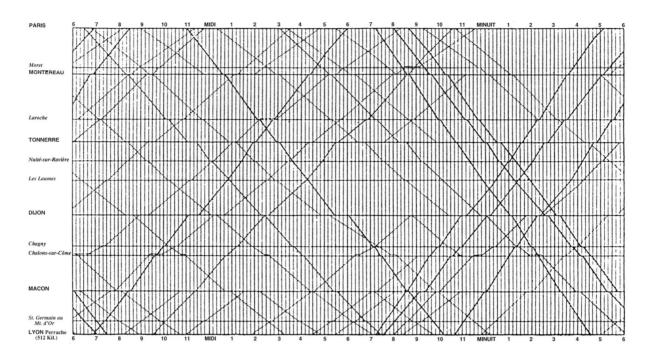

FIGURE 1. A nineteenth-century French train schedule showing all trains between Paris and Lyons. It was developed by the French engineer Ibry and published by E.J. Marey in 1885. (Reprinted with permission from E.R. Tufte. *The Visual Display of Quantitative Information.* Cheshire, CT: Graphics Press, 1983.)

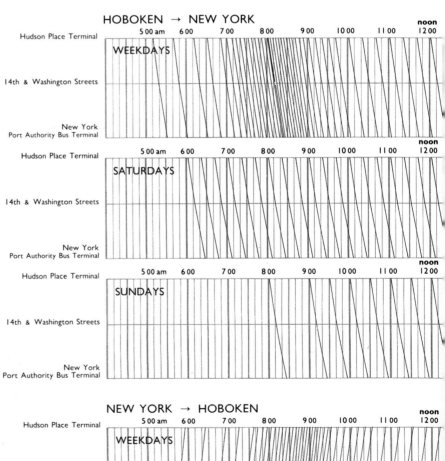

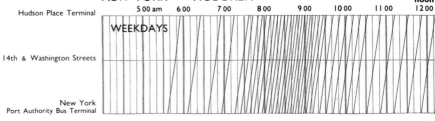

York City (figure 2). This schedule conveys many messages, three important ones are:

1. During rush hours, just show up.
2. During the rest of the day there is a bus every twenty minutes.
3. Between 1:15 and 5:00 A.M. forget it.

Both the topic and the title of this chapter were prompted by the management of the Radisson Hotel in Somerset, New Jersey, where Tufte gave his presentation. Guests who request it are given, free of charge, the Radisson's *Guest Information Guide.* It contains a copy of the schedule of trains that serve the Somerset area. Figure 3 is a carefully reproduced version of that timetable.

It tries to show forty-five trains and twenty two stations. Because of its initial smallish type and the resolution lost to multiple photocopy-

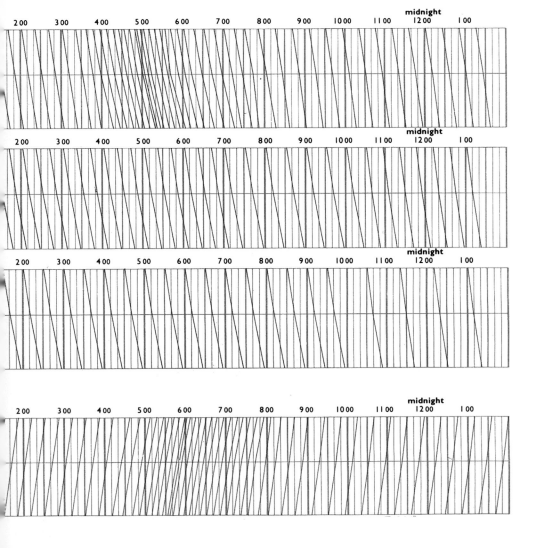

FIGURE 2. An excerpt from a modern version of Ibry's train schedule. The original shows all of the commuter buses between Hoboken and New York. Note that the time axis is divided into ten-minute segments to ease interpolation. (Reprinted with permission from a poster prepared by Edward Tufte, Inge Druckrey, and Nora Hillman Goeler and published by Graphics Press.)

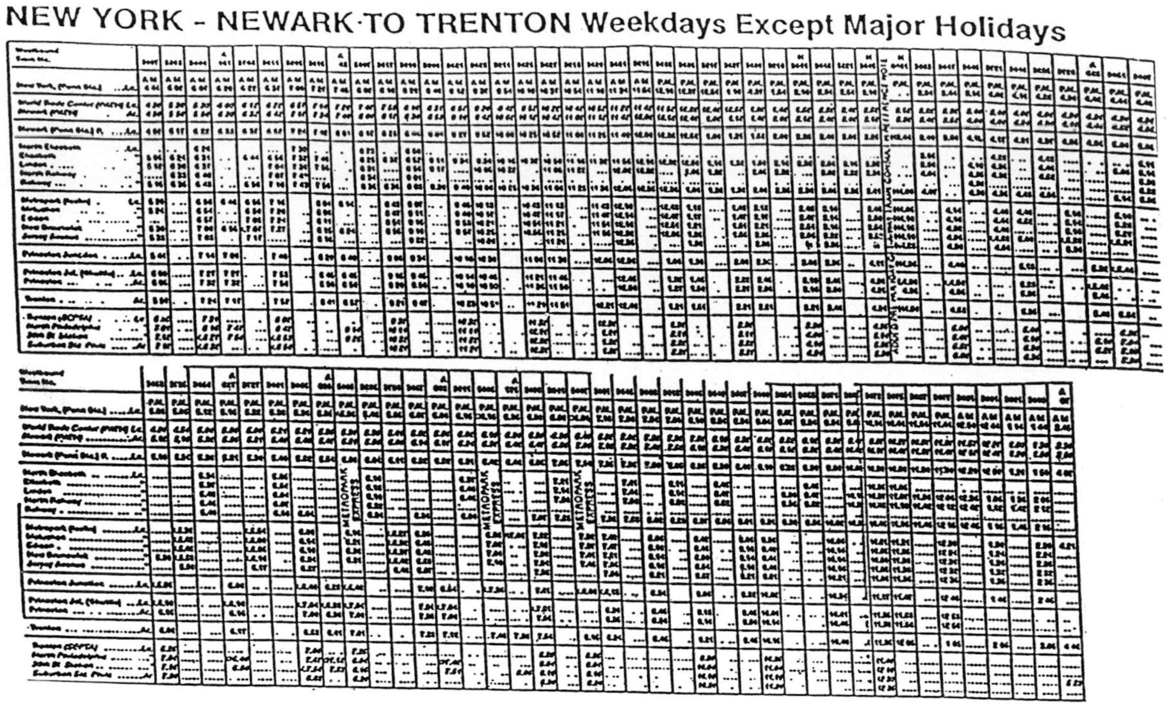

NEW YORK - NEWARK·TO TRENTON Weekdays Except Major Holidays

FIGURE 3. A good copy of the Trenton to New York train schedule supplied to guests by the Radisson Hotel in Somerset, New Jersey.

ing, it is absolutely unreadable. How ironic that in the same amount of space that was used to present an illegible tabular train schedule, careful design and production could have yielded a rich display providing us with useful, rememberable answers to questions both general and specific.

CHAPTER 7 Picturing an L.A. Bus Schedule

In 1991, Congress passed the National Literacy Act. The purpose of it is "to enhance the literacy and basic skills of adults, to ensure that all adults in the United States acquire the basic skills necessary to function effectively. . . in their work and in their lives. . . ." The National Adult Literacy Survey was instituted to provide Congress with information to assess both the state of adult literacy and, later, the success of programs instituted to affect it. The first National Adult Literacy Survey was conducted in 1992. Literacy within the survey was defined through a mixture of different kinds of problems that were thought to be real-world tasks.

It was generally felt by the sponsors of the National Adult Literacy Survey that an important aspect of being literate in modern society is having the knowledge and skills needed to process information from documents. In fact, for many adults, more time is spent reading tables, schedules, charts, graphs, maps, and forms of all kinds than any other type of material.[1] Thus, measuring document literacy was a major subtask of the survey.

Figure 1 shows one problem from this survey. It was meant to be on the rather more difficult end of the spectrum, because it required the understanding of a complex display with a lot of distracting information. The task was as follows: Using the bus schedule shown here, readers are asked how long they must wait for the next bus on a Saturday afternoon if they miss the 2:35 bus leaving Hancock and Buena Ventura going to Flintridge and Academy.

The reader is meant to note first that the proposed direction corresponds to the OUTBOUND direction. Next the reader must scan down the "Leave Hancock and Buena Ventura" column until reaching the 2:35 entry. Finally, one must notice that the next bus listed (3:05) does not run on Saturdays, so one must catch the bus that leaves at 3:35. Subtracting 2:35 from 3:35 yields a one-hour wait, which was scored as the correct answer. Of course, the question is ambiguous,

On Saturday afternoon, if you miss the 2:35 bus leaving Hancock and Buena Ventura going to Flintridge and Academy, how long will you have to wait for the next bus?

ROUTE 5

VISTA GRANDE

This bus line operates Monday through Saturday providing "local" service to most neighborhoods in the northeast section
Buses run thirty minutes apart during the morning and afternoon rush hours Monday through Friday
Buses run one hour apart at all other times of day and Saturday
No Sunday, holiday or night service.

You can transfer from this bus to another headed anywhere else in the city bus system

OUTBOUND (from Terminal)

Leave Downtown Terminal	Leave Hancock and Buena Ventura	Leave Citadel	Leave Rustic Hills	Leave North Carefree and Oro Blanco	Arrive Flintridge and Academy
AM					
6:20	6:35	6:45	6:50	7:03	7:15
6:50	7:05	7:15	7:20	7:33	7:45
7:20	7:35	7:45	7:50	8:03	8:15
7:50	8:05	8:15	8:20	8:33	8:45
8:20	8:35	8:45	8:50	9:03	9:15
8:50	9:05	9:15	9:20	9:33	9:45
9:20	9:35	9:45	9:50	10:03	10:15
10:20	10:35	10:45	10:50	11:03	11:15
11:20	11:35	11:45	11:50	12:03	12:15
PM					
12:20	12:35	12:45	12:50	1:03	1:15
1:20	1:35	1:45	1:50	2:03	2:15
2:20	2:35	2:45	2:50	3:03	3:15
2:50	3:05	3:15	3:20	3:33	3:45
3:20	3:35	3:45	3:50	4:03	4:15
3:50	4:05	4:15	4:20	4:33	4:45
4:20	4:35	4:45	4:50	5:03	5:15
4:50	5:05	5:15	5:20	5:33	5:45
5:20	5:35	5:45	5:50	6:03	6:15
5:50	6:05	6:15	6:20	6:33	6:45
6:20	6:35	6:45	6:50	7:03	7:15

INBOUND (toward Terminal)

Leave Flintridge and Academy	Leave North Carefree and Oro Blanco	Leave Rustic Hills	Leave Citadel	Leave Hancock and Buena Ventura	Arrive Downtown Terminal
AM					
6:15	6:27	6:42	6:47	6:57	7:15
6:45	6:57	7:12	7:17	7:27	7:45 Monday through Friday only
7:15	7:27	7:42	7:47	7:57	8:15
7:45	7:57	8:12	8:17	8:27	8:45 Monday through Friday only
8:15	8:27	8:42	8:47	8:57	9:15
8:45	8:57	9:12	9:17	9:27	9:45 Monday through Friday only
9:15	9:27	9:42	9:47	9:57	10:15
9:45	9:57	10:12	10:17	10:27	10:45 Monday through Friday only
10:15	10:27	10:42	10:47	10:57	11:15
11:15	11:27	11:42	11:47	11:57	12:15
12:15	12:27	12:42 p.m	12:47 p.m.	12:57 p.m.	1:15 p.m.
PM					
1:15	1:27	1:42	1:47	1:57	2:15
2:15	2:27	2:42	2:47	2:57	3:15
3:15	3:27	3:42	3:47	3:57	4:15
3:45	3:57	4:12	4:17	4:27	4:45 Monday through Friday only
4:15	4:27	4:42	4:47	4:57	5:15
4:45	4:57	5:12	5:17	5:27	5:45 Monday through Friday only
5:15	5:27	5:42	5:47	5:57	6:15
5:45	5:57	6:12	6:17	6:27	6:45 Monday through Friday only
					Monday through Friday only

To be sure of a smooth transfer tell the driver of this bus the name of the second bus you need

FIGURE 1. Test item from the 1992 National Adult Literacy Survey. (Taken from Kirsch, Jungeblut, Jenkins, and Kolstad, 1993, p. 91.)

yielding different answers depending how late you were for the 2:35. Someone who was fifty-five minutes late would hardly wait at all, whereas someone who missed the 2:35 by four and a half hours would have to wait almost two days. But pointing out flaws in the National Adult Literacy Survey is not my point. Instead, I would like to see what other kinds of questions could plausibly be asked of the data contained in the bus schedule and see whether another kind of display might facilitate the answering of those questions. I also note in passing that ignoring the ambiguity in the question, a person could answer the question without ever reading the contents of the bus schedule but merely by noting in the small print on top that "Buses run one hour apart (on) . . . Saturday."

My approach in this problem is to plot the bus data first and figure out what might be plausible questions later. At first this certainly seems a little backwards, but in fact, it is a reasonable strategy in what ought be an iterative process. Sometimes one has a data-related question and then draws a graph to try to answer it. After drawing the graph a new question might suggest itself, and hence a different graph, better suited to this new question (perhaps with additional data), is drawn. This in turn suggests something else, and so on, until either the data or the grapher is exhausted. With a circular process like this, there is no start or finish. One drops into the process anywhere and then proceeds. My experience suggests that if you begin with a general-purpose plot there is a greater chance of finding what you had not expected. So I had no hesitation, in this instance, in plotting the bus data first and using what I saw to direct further inquiries.

The next task then was deciding which general-purpose plot was most likely to allow unsuspected but interesting aspects of the data to show themselves. One natural way to do it, suggested by Marey's wonderful nineteenth-century French train schedule, is to plot time of the day on the horizontal axis and the various bus stops on the vertical axis. When I did this for a selected portion of the data and represented each bus as a line on this plot, the display shown in figure 2 resulted.

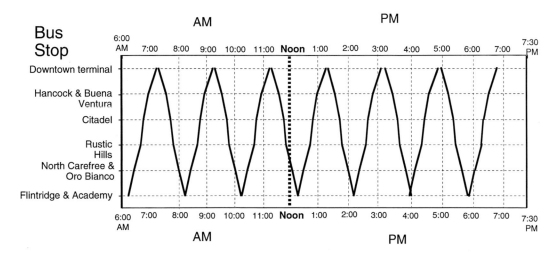

This display told me two things that perhaps I should have suspected, but which are now obvious. First, and most important, the cyclic nature of the graph strongly suggests that a single bus is going back and forth on this route, making a round trip in just under two hours (and six and a half round trips each day). Second, the lines drawn representing the buses' paths are irregular, suggesting either that buses vary in speeds over their routes or that the bus stops are irregularly spaced. The latter hypothesis is surely the more useful, so I straightened the lines by adjusting the spacing between the stops. The next iteration reconfigures the stops so that the bus paths are straight lines and all the data are included. Figure 3 is the result. As is now obvious, the Citadel and Rustic Hills stops are much closer together than all the rest, which remain spaced about equally.

FIGURE 2. Selected data from figure 1 plotted in a graphical form suggested by Marey (1885) and described in chapter 6.

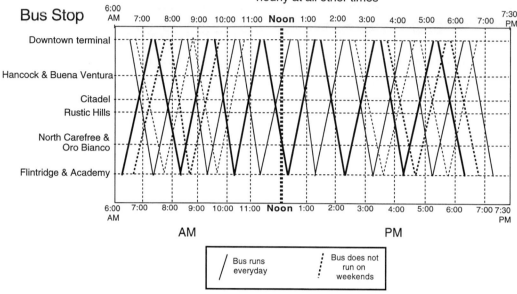

FIGURE 3. A redraft of figure 2 in which the bus stops are spaced proportionally to the time taken for a bus to go between them and including all data from figure 1.

I noticed that there appear to be only two buses assigned to this route six days a week but that two others also ply these streets during the morning and afternoon periods of heaviest traffic during the week. I chose to represent each of these four buses with a different line type, dashing the lines representing the two buses that did not run on Saturdays.

What are the uses for this version of the bus schedule? Certainly, from the point of view of a bus company official, the graphical schedule shown in figure 3 is a superior management tool. It tells at a glance how many buses are on the route, where and when drivers must be changed, when drivers can take breaks, and so forth. One can also imagine a larger display, made up of many panels, in which each panel is another route. Such a display would help to show where the two extra rush hour buses are coming from, what they do when they are not on this route, and so on. Another alternative would be a much more extensive version including all Los Angeles bus routes. The extent to which this is possible is not known without actually producing such a schedule, but such displays are used as a matter of course by controllers of the Tokyo mass transit system, and there is no reason to believe that the Los Angeles bus system is more complex. Of course, if a particular display is expected to be used primarily as a management tool, it can become an implicit display, in which all of the routes are in the machine but only a subset is displayed, as they are requested (*omnibus ex machina?*).

But what about the person for whom the printed bus schedule is principally intended, the poor straphanger? I certainly believe that a rider can use the system better if fully informed about the overall structure of service. Being so informed is easier from figure 3 than from figure 1. But what about the extraction of small bits of information? What about answering the question posed in the National Adult Literacy Survey? Suppose we remove the ambiguity in the question asked by amending it to be, As you arrive one Saturday afternoon, at the corner of Hancock and Buena Ventura, you catch a glimpse of the 2:35 P.M. bus to Flintridge and Academy heading off to its next stop. How long will it be before the next bus to Flintridge and Academy arrives? The answer to this question can be found in figure 3, although not to the precision of that shown in the original tabular schedule. Shown in figure 4 is an enlargement of the relevant portion of figure 3 with the missed bus shown as a white circle and the next one as a darkened one. The hour difference between them is clear. But of course, if we knew the general structure of all bus routes, we would know that there is always an hour between buses on Saturdays and a half hour between them during the weekday rush hours.

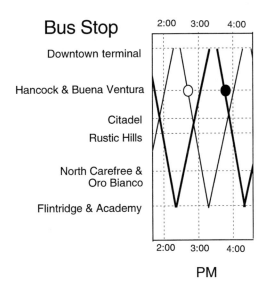

Route 5 Vista Grande
(Saturday afternoon)

FIGURE 4. The most relevant section of figure 3 shown expanded with the bus stop of interest from the survey question indicated.

Figure 4 was built especially for the question asked. An even better special-purpose display might be simply "one hour to wait"—obviously a very special-purpose display. In fact, it is so special that even if one could carry around some sort of small electronic device that had the entire bus schedule stored in its innards, it might take more time and effort to input the question than would be taken by scanning a more general display. Indeed, it is easy to imagine a general-purpose device that might have (among many other things) all of the Los Angeles bus routes inside, displayed in some large general format. Thus, with some kind of pan and zoom mechanism built in, it would allow the user to zero in on an answer as precisely as needed. This sort of technology is already widely available for laptop computers (e.g., *Streetmap*™, which allows one to look at maps as general as the entire United States and zoom in to be as specific as the street on which you live, the latter at a scale enlarged enough to locate house numbers). I see no reason why *Streetmap*™-like software won't become available eventually for cheap pocket computers of the sort now called "personal organizers."

Obviously, I am a fan of graphic schedules, but I am not offering figure 3 as the final answer. To determine what sort of display works best would require some experimentation. It may be that legibility is increased by separating weekday and Saturday displays and/or separating ingoing and outgoing buses onto separate panels as well. Edward Tufte used this latter approach in his redesigned PATH bus schedules for the Hoboken-New York run.* Splitting the data simplified the picture but removed the possibility of seeing the total path of an individual bus, admittedly a goal of only limited interest to a bus passenger.

It would have been interesting to have included some version of figure 3 on the National Adult Literacy Survey and so to be able to compare people's performance on it with their performance on the traditional display. But the goal of the survey was to see how people perform on what is "out there," not how their performance might be improved by changing it. Maybe next time.

*One of which was shown as figure 2 in the chapter 6.

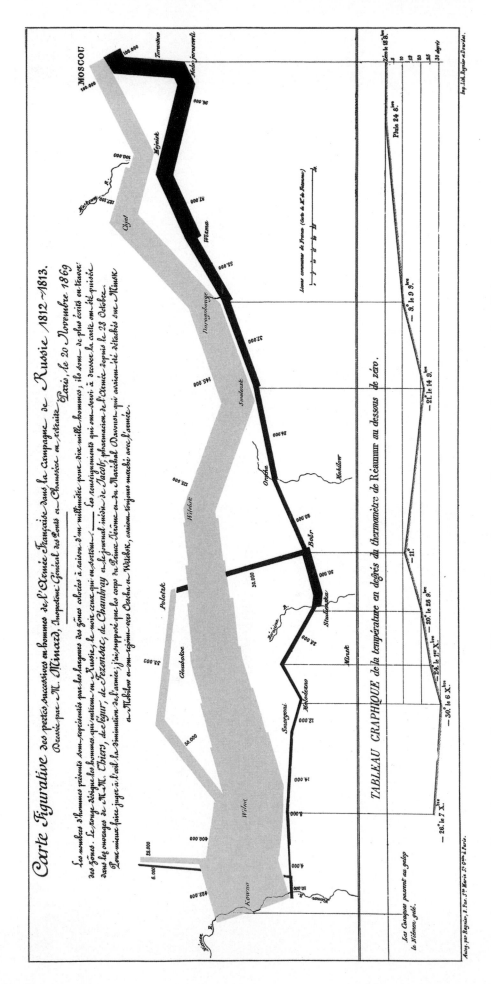

CHAPTER 4, FIGURE 1. Charles Joseph Minard's famous plot of Napoleon's Russian Campaign. (Reprinted, with permission, from Tufte, 1983.)

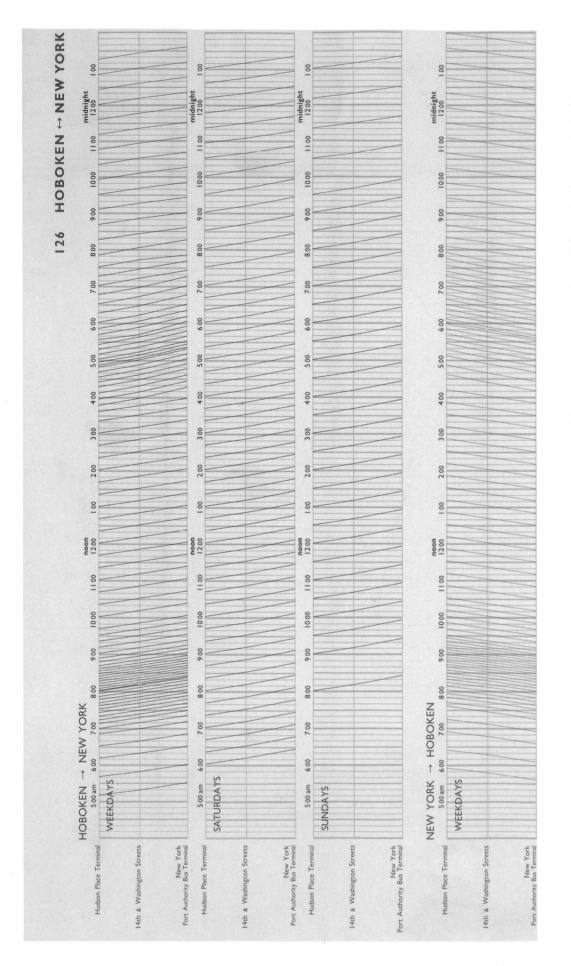

CHAPTER 6, FIGURE 2. An excerpt from a modern version of Ibry's train schedule prepared by Edward Tufte, Inge Druckrey, and Nora Hillman Goeler and published by Graphics Press. The original shows all of the commuter buses between Hoboken and New York. Note that the time axis is divided into ten-minute segments to ease interpolation. (Reprinted with permission from a poster prepared by Edward Tufte, Inge Druckrey, and Nora Hillman Goeler and published by Graphics Press.)

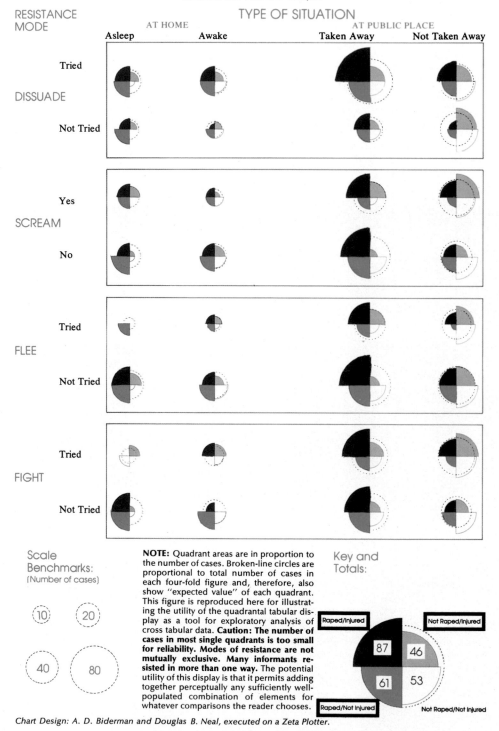

Frequencies of Rapes and Injuries in Various Types of Stranger Rape Situations by Whether or Not a Mode of Resistance Was Used

Quadrantal Tabular Graph

Scale Benchmarks: (Number of cases)

10 20 40 80

NOTE: Quadrant areas are in proportion to the number of cases. Broken-line circles are proportional to total number of cases in each four-fold figure and, therefore, also show "expected value" of each quadrant. This figure is reproduced here for illustrating the utility of the quadrantal tabular display as a tool for exploratory analysis of cross tabular data. **Caution: The number of cases in most single quadrants is too small for reliability. Modes of resistance are not mutually exclusive. Many informants resisted in more than one way.** The potential utility of this display is that it permits adding together perceptually any sufficiently well-populated combination of elements for whatever comparisons the reader chooses.

Key and Totals:

Raped/Injured 87
Not Raped/Injured 46
Raped/Not Injured 61
Not Raped/Not Injured 53

Chart Design: A. D. Biderman and Douglas B. Neal, executed on a Zeta Plotter.

CHAPTER 11, FIGURE 6. A multivariate application of the two-way rose prepared by Albert Biderman and Douglas Neal in 1979. (Reprinted from the spring-summer 1979, vol. XIII, 2–3, newsletter of the Bureau of Social Science Research, Washington, D.C.)

CHAPTER 11, FIGURE 8. A more literal icon used to show wind direction from Juan de La Cosa's 1500 map of the New World. The map is notable for its depiction of Cuba as an island. In addition, could this "wind gnome" be the eponymous source for the term "gnomon" that is used to describe the gadget that shows the direction of the light on sun dials?

Graphical Forms

Within the mass media the only explanation that makes any sense for the way that graphics are used is the power of convention. Pie charts are but one example of many graphical forms that are used broadly but are generally poor choices for effective display. In contrast, there are other display forms, like the Nightingale rose, that are likely to be of enormous usefulness yet are rarely found. Why? A full explanation is complex, but certainly part of the answer lies in convention. Once we have learned and internalized a particular form, it allows us to absorb (or at least think that we have absorbed) the information within it with little cognitive processing. As in many other situations, whatever gets there first has an advantage. The irony is that as graphical forms evolve and improve, the success of earlier formats works against the easy adoption of the newer forms.

This section divides neatly in half. Each of the first three chapters describes a single commonly used presentation format, points out its shortcomings, and suggests either improvements, alternatives, or limitations on the areas of its usefulness. The next three chapters describe three much more rarely used formats and show how they can provide us with deep insights into certain kinds of data. The obstacle that these three forms face is their unfamiliarity. I have no doubt that but for the power of convention they would be as familiar as they are beautiful.

Chapter 8 discusses pie charts. Pie charts are the most familiar graphical form, and, surprisingly, the least useful. Anyone doubting this should try to display more than three or four numbers unambigu-

ously with a pie. In this chapter I show some misuses of the pie chart and some attractive alternatives.

Chapter 9 describes the most easily misused graphical form ever invented, the double Y-axis plot ("plot" is used here in its most pejorative connotation). I used such a plot in the introduction to section I to show how one could shift the emphasis of the principal finding of the Surgeon General's report on smoking away from tobacco as the cause of increased mortality. The double Y-axis graph is commonly used to compare two variables that are on very different scales. Since the point of the format is to maintain the differences in two scales, "comparison" is exactly the kind of inference that is illegitimate. Pie charts are of limited value because they hide what we might have seen in a more transparent format; double Y-axis graphs, because they mislead us, are much worse.

Chapter 10 is about tables. A more boring topic is hard to imagine. Yet much of the quantitative information we encounter is likely to be in tabular form; scan the sports or business pages of the newspaper or the information booklets from the IRS to confirm this observation. It was astonishing to me to see how well tables can work in communicating complex data when they are thoughtfully prepared. In chapter 10 I present and illustrate a few simple rules that if followed yield that apparent oxymoron, interesting tables.

Pie charts are old; William Playfair invented them in the late eighteenth century. But in 1857, when Florence Nightingale used another kind of circular plot to show the relationship between weather and mortality in the Crimean War, she was using a graph with a 400-year history. In chapter 11 the origins of the beautiful Nightingale rose is traced back to the fifteenth century and its effective use is illustrated on some modern data.

Chapter 12 tells of a somewhat more esoteric graph, the trilinear plot. It is a graphic that takes some getting used to, but when it is suitable it can provide us with a clearer view of complex data than any other form. Upton's trilinear plots of the British election results are but one compelling example.

Completing this section, chapter 13 tells of a graphic that is all form with no content—an implicit graph. The user fills in the numbers. As one example, I have used an entire page of this book to reproduce a piece of constant-dollar graph paper. Plot your salary on it and be prepared to grieve.

CHAPTER 8 Humble Pie

I used to like pie charts, but that was a long time ago. Now I hate them. What brought about this change of heart? Reasons for hating pie charts are so numerous that almost no explanation is necessary. What is more surprising is why I liked them originally.

Many years ago I observed that "graphicacy" in children preceded "numeracy"; a child could recognize that one third was larger than one fourth from a pie chart long before being able to make that determination from the numerical representations of the fractions themselves. I viewed this (erroneously) as a tribute to pie charts. Actually it is a tribute to the powerful graphical sense that humans have. We can make correct judgments about data even if they are pied!

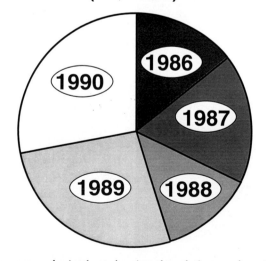

**Number of International Flights
from Newark Airport
(1986-1990)**

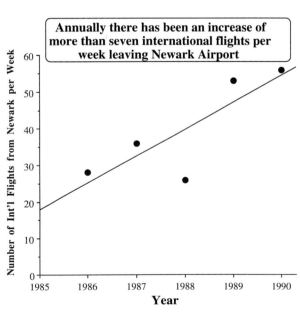

FIGURE 1. A pie chart showing the relative number of international flights from Newark Airport over a five-year period. The graphics software didn't squawk at all when asked to make a pie chart with such inappropriate content.

FIGURE 2. A more informative plot of the same flight data shown previously in figure 1. Now the size and direction of the trend are obvious.

Modern software for computer graphics can make a pie out of anything. Consider the chart shown in figure 1. A more appropriate version is shown in figure 2.

But you say: Aren't there rules about when and how to use pie charts? Yes, indeed. Let me quote one authoritative source, the *Publication Manual of the American Psychological Association:*

> Circle (or pie) graphs, by nature 100% graphs, are used to show percentages and proportions. The number of items compared should be kept to five or fewer. Order the segments from large to small, beginning the largest segment at 12 o'clock. A good way to highlight differences is to shade the segments from light to dark, making the smallest segment the darkest. Use patterns of lines and dots to shade the segments.[1]

FIGURE 3. A pie chart with remarkably apt content. (© 1990 The New York Times Company. Reprinted with permission.)

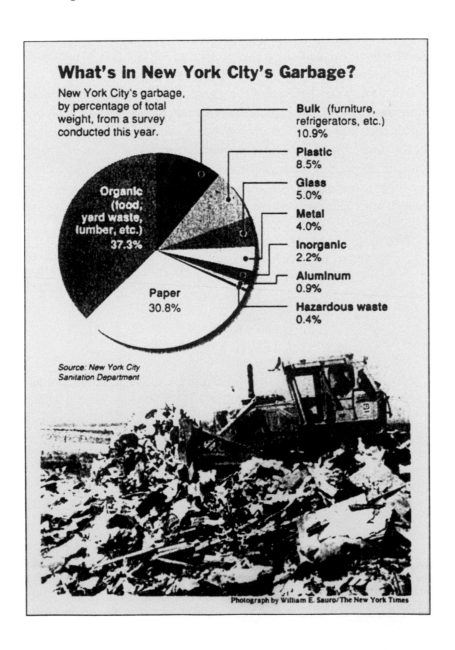

What's in New York City's Garbage?

New York City's garbage, by percentage of total weight, from a survey conducted this year.

Organic (food, yard waste, lumber, etc.) 37.3%

Paper 30.8%

Bulk (furniture, refrigerators, etc.) 10.9%

Plastic 8.5%

Glass 5.0%

Metal 4.0%

Inorganic 2.2%

Aluminum 0.9%

Hazardous waste 0.4%

Source: New York City Sanitation Department

Photograph by William E. Sauro/The New York Times

Graphics software that I'm familiar with does not help you decide whether a particular data set is appropriate for a pie; however, it generally does follow most of the above rules (except for shading). In figure 3 is a pie chart with an extraordinarily appropriate content. It docs have more than the recommended five segments, but they are ordered and are carefully labeled. Even though some segments are inaccurately portrayed (compare the segments for metal vs. glass), this is an example of a pie chart done more or less properly.

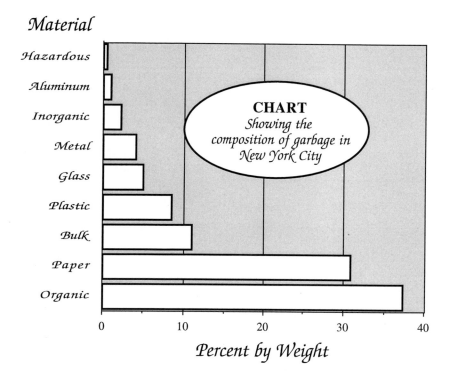

FIGURE 4. The data from figure 3 redisplayed as a horizontal bar chart.

An alternative bar chart is shown in figure 4. I contend that the relative sizes of the various components of garbage are clearer. There is room for more detail if it is needed. While it is true that the sum of all the components is not as clearly shown to be 100% as with a pie, that is a small price to pay for the increased clarity of the data depiction. Note that the format and fonts used are a little old-fashioned. This is my attempt to retrieve some of the elegance in displays that William Playfair so ably demonstrated, but that is the grist for chapter 18.

My last example (figure 5), from the *Ottawa Citizen*, is more typical. It violates the standards for pie charts. There are too many segments, and the area of each segment is unrelated to the number of dollars that it contains. But even if the standards were followed it would still be opaque. Redoing this pie as an ordered bar chart (figure 6) provides a display that actually communicates the data. I have yet to find a pie chart that cannot be improved in this way.

FIGURE 5. Another pie of limited value from the *Ottawa Citizen* February 9, 1989, (Reprinted with permission from Rob Cross.)

FIGURE 6. The data from figure 5 redisplayed as a horizontal bar chart

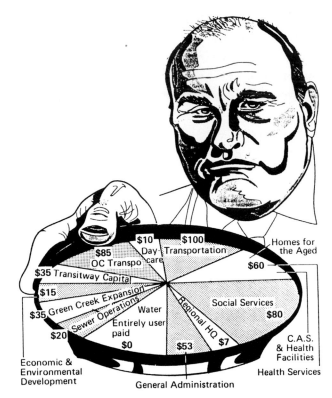

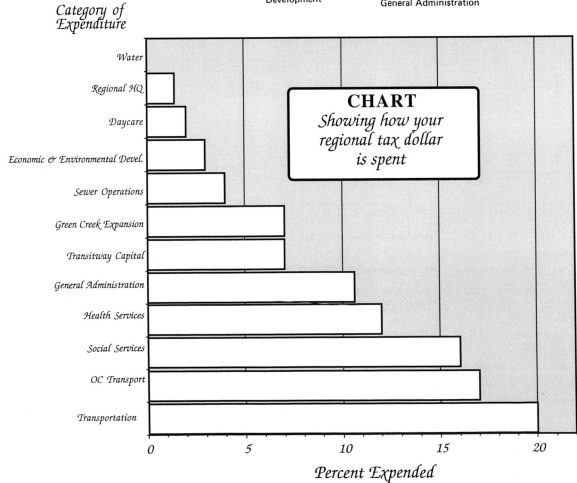

Category of Expenditure

Percent Expended

CHAPTER 9 Double Y-Axis Graphs

The use of a pie chart is sinful, but the sin is venial. The sin of a "double Y-axis graph" is mortal. If there is a just God, I am sure that there is a special place in the inferno reserved for its perpetrators.

A *double Y-axis graph* is a format that refers one function to the left axis and a second function to the right. Most graphics packages offer this type of graph as an easily used option. It allows one to easily twist the facts to suit one's aims.

Figure 1 was seen earlier in the introduction to section I. It provides a rough duplicate of a plot that appeared in the 1964 Surgeon General's report *Smoking and Health*. The legend has been modified to be more informative.

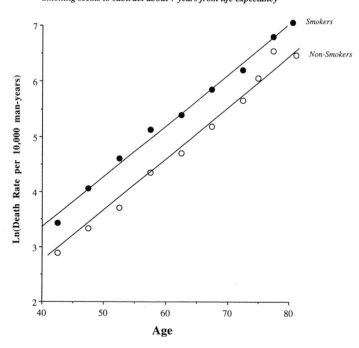

FIGURE 1. A (very slightly) redone version of a figure from the 1964 Surgeon General's report *Smoking and Health*.

One could, puckishly, consider what the reaction to such a graph might be from defenders of the tobacco industry. I imagine that a statistician working in this industry who brought such a figure to his boss as an example of clear data display might be asked to rework the plot. The double Y-axis format is just the thing for obscuring these data. In figure 2 is such a plot (worthy of the name) with a suitably informative legend.

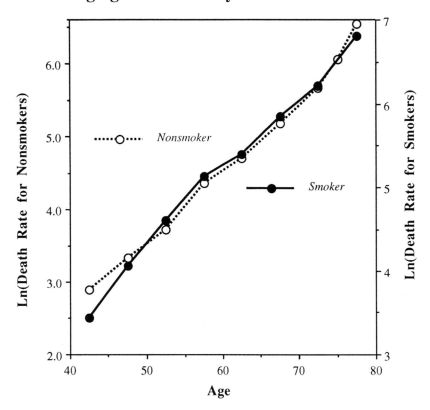

FIGURE 2. Transforming the data from figure 1 into a double Y-axis plot effectively hides the effect smoking has on mortality.

Of course this transformation of an informative graph into one that misinforms is, to the best of my knowledge, hypothetical. Is this potentially confusing format actually used? You betcha!

Figure 3 is taken from the May 14, 1990 issue of *Forbes* magazine. It purports to show that while per pupil expenditures for education have gone up precipitously over the last decade, student performance (as measured by mean SAT scores) has not responded. The conclusion is that we ought not to waste our money on education.

Of course, basing such inferences on these data is completely specious. Since we have complete control of both Y-axes, we can rescale to yield any result we want. Consider the alternatives shown in figures 4 and 5.

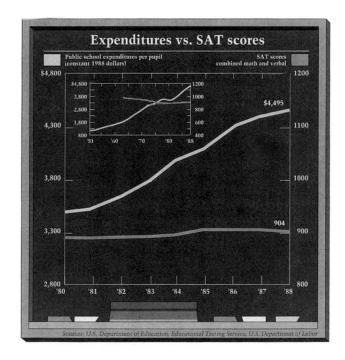

FIGURE 3. A double Y-axis graph from *Forbes* magazine, May 14, 1990 (© Forbes Inc., 1990) apparently showing that increased spending on education has had no effect on SAT scores.

In figure 4 we scale both Y-axes to cover the range of the data (just as was done in figure 2). This yields the conclusion that both variables move apace.

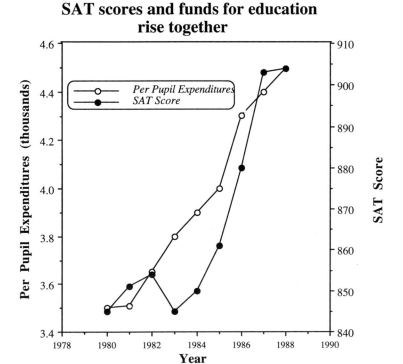

FIGURE 4. The data from figure 3 redisplayed when both vertical axes are scaled to expose the changes maximally.

In figure 5 we stretch the expenditure axis from $0 to $20,000, thus making the educational outlays appear penurious. At the same time, we expand the SAT axis so that the series begins above the expenditure series and appears to increase massively.

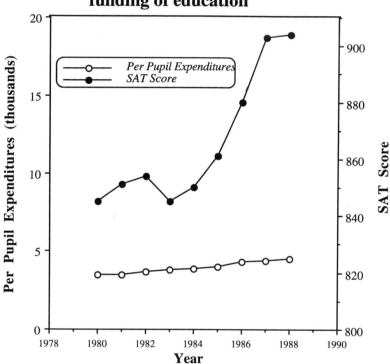

FIGURE 5. The data from figure 3 redisplayed with both vertical axes scaled to suggest that SAT scores have increased despite sluggish spending on education.

Is there any use of the double Y-axis format that is reasonable? A very restricted "yes." Sometimes the same dependent variable can be represented in a transformed way. For example, plot *log of per pupil expenditures* on the left and *per pupil expenditures* on the right, the latter spaced to match the left-hand scale. Thus, those for whom *log dollars* isn't helpful can look on the other scale. Similarly, plots of population size against age can be augmented with an axis parallel to the age axis labeled by year of birth. Ironically, no graphics package I know of allows this latter use to be done easily, whereas the misuse is often a touted option. Alas!

CHAPTER 10 Tabular Presentation

Getting information from a table is like extracting sunlight from a cucumber.

—A.B. Farquhar and H. Farquhar[1]

The disdain shown by the two nineteenth-century economists quoted above reflected a minority opinion at that time. Since then the use of graphs for data analysis and communication has increased but their quality has, in general, deteriorated since Playfair's death. Tables, spoken of so disparagingly by the Farquhars, remain, to a large extent, worthy of contempt. My primary focus in this chapter is the improvement of tabular presentation. Toward this end I will discuss and illustrate three simple rules for the preparation of useful tables.

Driving these rules is the orienting attitude that a table is for communication, not data storage. Modern data storage is accomplished well on magnetic disks or tapes, optical disks, or some other mechanical device. Paper and print are meant for human eyes and human minds.

We begin with table 1, table 5/19 in the Bureau of the Census's well-known book *Social Indicators III*.

Any redesign task must first try to develop an understanding of purpose. The presentation of this data set must have been intended to help the reader answer such questions as:

1. What is the general level (per 100,000 population) of accidental death in the countries chosen?
2. How do the countries differ with respect to their respective rates of accidental death?
3. What are the principal causes of accidental death? Which are the most frequent? The least frequent?
4. Are there any unusual interactions between country and cause of accidental death?

TABLE 1

Deaths Due to Unexpected Events, by Type of Event, Selected Countries: Mid-1970's

(Rate per 100,000 population)

Country	Year[1]	Deaths due to all causes	Deaths due to unexpected events					
			Total	Transport accidents	Natural factors[2]	Accidents occurring mainly in industry[3]	Homicides and injuries caused intentionally[4]	Other causes[5]
Austria..............	1975	1,277.2	75.2	34.8	29.7	4.3	1.6	4.8
Belgium.............	1975	1,218.5	62.6	25.0	25.8	1.5	9	9.4
Canada.............	1974	742.0	62.1	30.9	18.0	3.9	2.5	6.8
Denmark............	1976	1,059.5	41.1	18.3	15.6	1.0	7	5.5
Finland.............	1974	952.5	62.3	23.7	26.0	2.9	2.6	7.1
France.............	1974	1,049.5	77.8	23.8	31.0	1.0	9	21.1
Germany (Fed. Rep.)..	1975	1,211.8	66.4	24.8	31.6	1.8	1.2	7.0
Ireland.............	1975	1,060.7	48.6	19.8	20.1	1.9	1.0	5.8
Italy...............	1974	957.8	47.2	22.8	19.2	1.9	1.1	2.2
Japan..............	1976	625.6	30.5	13.2	9.7	2.1	1.3	4.2
Netherlands........	1975	832.2	40.3	17.8	18.2	1.0	7	2.6
Norway.............	1976	998.9	48.4	17.3	25.1	1.9	7	3.4
Sweden............	1975	1,076.6	55.8	17.2	27.9	1.3	1.1	8.3
Switzerland........	1976	904.1	48.4	20.6	20.4	2.1	9	4.4
United Kingdom......	1976	1,217.9	34.8	13.0	13.9	1.3	1.1	5.5
United States.......	1975	888.5	60.6	23.4	15.8	2.6	10.0	8.8

[1]Most current year data available.
[2]Includes fatal accidents due to poisoning, falls, fire, and drowning.
[3]For some countries data relate to accidents caused by machines only.
[4]By another person, including police.
[5]Includes accidents caused by firearms, war injuries, injuries of undetermined causes, and all other accidental causes.

Source: United Nations, World Health Organization, World Health Statistics Annual, 1978, vol. I, Vital Statistics and Cause of Death. Copyright; used by permission.

These are obviously parallel to the questions that are ordinarily addressed in the analysis of any multifactorial table—overall level, row, column, and interaction effects.

Before going further I invite you to read table 1 carefully and see to what extent you can answer these four questions. But don't peek ahead!

The first rule of table construction is:

I.   Order the rows and columns in a way that makes sense.

We are almost never interested in "Austria First." Two useful ways to order the data are:

i.   Size places—Put the largest first. Often we look most carefully at what is on top and less carefully further down. Put the biggest thing first! Also, ordering by some aspect of the data often reflects ordering by some hidden variable that can be inferred.

ii.   Naturally—Time is ordered from the past to the future. Showing data in that order melds well with what the viewer might expect. This is always a good idea.

Table 2 is a redone version of table 1. A few typos have been corrected, some uninformative columns removed, and the rows ordered by the total death rate. The columns were already ordered in a reasonable way and so were left unaltered. Now we can begin to

TABLE 2

**Deaths Due to Unexpected Events, by Type of Event,
Selected Countries: Mid-1970s**
(Rate per 100,000 population)

Country	Total unexpected deaths	Transport accidents	Natural factors	Industrial accidents	Homicides	Other causes
France	77.8	23.8	31.0	1.0	0.9	21.1
Austria	75.2	34.8	29.7	4.3	1.6	4.8
Germany	66.4	24.8	31.6	1.8	1.2	7.0
Belgium	62.6	25.0	25.8	1.5	0.9	9.4
Finland	62.3	23.7	26.0	2.9	2.6	7.1
Canada	62.1	30.9	18.0	3.9	2.5	6.8
United States	60.6	23.4	15.8	2.6	10.0	8.8
Sweden	55.8	17.2	27.9	1.3	1.1	8.3
Ireland	48.6	19.8	20.1	1.9	1.0	5.8
Norway	48.4	17.3	25.1	1.9	0.7	3.4
Switzerland	48.4	20.6	20.4	2.1	0.9	4.4
Italy	47.2	22.8	19.2	1.9	1.1	2.2
Denmark	41.1	18.3	15.6	1.0	0.7	5.5
Netherlands	40.3	17.8	18.2	1.0	0.7	2.6
United Kingdom	34.8	13.0	13.9	1.3	1.1	5.5
Japan	30.5	13.2	9.7	2.1	1.3	4.2

answer questions 1 and 2 above. We see that France is the most dangerous place, having an accidental death rate of about seventy-eight per hundred thousand, that is, more than twice that of Japan (about thirty per hundred thousand), which, at least by this measure, appears to be the safest country. Now that the rows are ordered, the overall death rate (taken as an unweighted median) can be easily calculated—count down eight countries—at around fifty per hundred thousand.

Note that when I referred to the actual rates, I rounded. This is very important. The second rule of table construction is:

II.   Round—a lot!

This is for three reasons:

   i.   Humans cannot understand more than two digits very easily.
   ii. We can almost never justify more than two digits of accuracy statistically.
   iii. We almost never care about accuracy of more than two digits.

Let us take each of these reasons separately.

Understanding. Consider the statement, "This year's school budget is $27,329,681." Who can comprehend or remember that? If we remember anything, it is almost surely the translation, "This year's school budget is about 27 million dollars."

Statistical justification. The standard error of any statistic is proportional to one over the square root of the sample size. God did this, and there is nothing we can do to change it. Thus, suppose we would like to report some statistic (like a correlation) whose estimated value is 0.25. If we don't want to report something that is inaccurate, we must be sure that the second digit is reasonably likely to be 5 and not 6 or 4. To accomplish this we need the standard error to be less than 0.005. But since the standard error is proportional to $1/\sqrt{n}$, the obvious algebra $(1/\sqrt{n} \sim 0.005 \Rightarrow \sqrt{n} \sim 1/0.005 = 200)$ yields the inexorable conclusion that a sample size of the order of 200^2, or 40,000, is required to justify the presentation of a two-digit correlation. A similar argument can be made for most other statistics.

Who cares? Table 1 of chapter 1 (p. 37) proudly reported the mean life expectancy of a male at birth in Spain was 58.76 years. What does the "6" mean? Each unit in the hundredths digit of this overzealous reportage represents four days. What purpose is served in knowing a life expectancy to this accuracy? For most communicative (not archival) purposes "59" would have been enough.

TABLE 3

**Deaths Due to Unexpected Events, by Type of Event,
Selected Countries: Mid-1970s**
(Rate per 100,000 population)

Country	Total unexpected deaths	Transport accidents	Natural factors	Industrial accidents	Homicides	Other causes
France	78	24	31	1	1	21
Austria	75	35	30	4	2	5
Germany	66	25	32	2	1	7
Belgium	63	25	26	2	1	9
Finland	62	24	26	3	3	7
Canada	62	31	18	4	3	7
United States	61	23	16	3	10	9
Sweden	56	17	28	1	1	8
Ireland	49	20	20	2	1	6
Norway	48	17	25	2	1	3
Switzerland	48	21	20	2	1	4
Italy	47	23	19	2	1	2
Denmark	41	18	16	1	1	6
Netherlands	40	18	18	1	1	3
United Kingdom	35	13	14	1	1	6
Japan	31	13	10	2	1	4

Table 3 contains a revision of table 2 in which each entry is rounded to the nearest integer. Because the original entries had only one extra digit, the clarifying effect of rounding is modest. In this version of the table the unusual homicide rate of the United States jumps out at us. At a glance we can see that it is an order of magnitude greater than that found in any civilized nation. We also see an unusual entry for France under "other causes" that raises questions about definitions.

The effect of too many decimal places is sufficiently pernicious that I would like to emphasize the importance of rounding with another short example. Equation (1) is taken from a book called *State Court Caseload Statistics: 1976*.

$$\text{Ln(DIAC)} = -10729131 + 1.00716993 \times \text{Ln(FIAC)}, \qquad (1)$$

where DIAC is the annual number of case dispositions and FIAC is the annual number of case filings. This is obviously the result of a statistical (regression) analysis with an overgenerous output format. Using the standard error justification for rounding we see that to justify the eight digits shown we would need a standard error that is of the order of 0.000000005, or a sample size of the order of 4×10^{16}. This is a very large number of cases—the population of China doesn't put a dent in it. The actual n is the number of states, which allows one digit of accuracy at most. If we round to one digit and transform out of the log metric, we arrive at the more statistically defensible equation

$$\text{DIAC} = 0.9 \text{ FIAC}. \qquad (2)$$

This can be translated into English as

THERE ARE ABOUT 90% AS MANY DISPOSITIONS AS FILINGS.

Obviously the equation that is more defensible statistically is also much easier to understand. My colleague Al Biderman, who knows more about courts than I do, suggested that I needed to round further, to the nearest integer (DIAC = FIAC), and so a more correct statement would be

THERE ARE ABOUT AS MANY DISPOSITIONS AS FILINGS.

A minute's thought about the court process reminds one that it is a pipeline with filings at one end and dispositions at the other. They must equal one another, and any variation in annual statistics reflects only the vagaries of the calendar. The sort of numerical sophistry demonstrated in equation 1 can give statisticians a bad name.*

III. ALL is different and important.

Summaries of rows and columns are important as a standard for comparison—they provide a measure of usualness. The type of summary

*I sometimes hear from colleagues that my ideas about rounding are too radical. That such extreme rounding would be "OK if we know that a particular result was final. But our final results may be used by someone else as intermediate in further calculations. Too early rounding would result in unnecessary propagation of error." Keep in mind that tables are for communication, not archiving. Round the numbers, and if you must, insert a footnote proclaiming that the unrounded details are available from the author. Then sit back and wait for the deluge of requests.

we use to characterize ALL depends on the purpose. Sometimes a sum is suitable, more often a median. But whichever is chosen, it should be visually different from the individual entries and set spatially apart.

TABLE 4

Deaths Due to Unexpected Events, by Type of Event, Selected Countries: Mid-1970s
(Rate per 100,000 population)

Country	Total unexpected deaths	Transport accidents	Natural factors	Industrial accidents	Homicides	Other causes
France	78	24	31	1	1	21
Austria	75	35	30	4	2	5
Germany	66	25	32	2	1	7
Belgium	63	25	26	2	1	9
Finland	62	24	26	3	3	7
Canada	62	31	18	4	3	7
United States	61	23	16	3	10	9
Sweden	56	17	28	1	1	8
Ireland	49	20	20	2	1	6
Norway	48	17	25	2	1	3
Switzerland	48	21	20	2	1	4
Italy	47	23	19	2	1	2
Denmark	41	18	16	1	1	6
Netherlands	40	18	18	1	1	3
United Kingdom	35	13	14	1	1	6
Japan	31	13	10	2	1	4
Median	**53**	**22**	**20**	**2**	**1**	**6**

Table 4 makes it clearer how unusual the United States homicide rate is. The column medians allow us to compare the relative danger of the various factors. We note that although "transport accidents" is the worst threat, it is closely followed by "natural factors." Looking at the entries for the United States, we can see that "natural factors" are under somewhat better control than in most other countries.

Can we go further? Sure. To see how requires that we consider what distinguishes a table from a graph. A graph uses space to convey information. A table uses a specific iconic representation. We have made tables more understandable by using space—making a table more like a graph. We can improve tables further by making them more graphical still. A semigraphical display like the stem-and-leaf diagram[2] is merely a table in which the entries are not only ordered but are also spaced according to their size. To put this notion into practice, consider the last version of table 1 shown as table 5.

TABLE 5

**Deaths Due to Unexpected Events, by Type of Event,
Selected Countries: Mid-1970s**
(Rate per 100,000 population)

Country	**Total unexpected deaths**	Transport accidents	Natural factors	Industrial accidents	Homicides	Other causes
France	78	24	31	1	1	21 +
Austria	75	35 +	30	4	2	5
Germany	66	25	32	2	1	7
Belgium	63	25	26	2	1	9
Finland	62	24	26	3	3	7
Canada	62	31 +	16 -	4	3	7
United States	61	23	18 -	3	10 +	9
Sweden	56	17	28	1	1	8
Ireland	49	20	20	2	1	6
Norway	48	17	25	2	1	3
Switzerland	48	21	20	2	1	4
Italy	47	23	19	2	1	2
Denmark	41	18	16	1	1	6
Netherlands	40	18	18	1	1	3
United Kingdom	35	13	14	1	1	6
Japan	31	13	10	2	1	4
Median	53	22	20	2	1	6

+ = an unusually high data value.

- = an unusually low data value.

The rows have been spaced according to what appear to be significant gaps[3] in the total death rate, dividing the countries into five groups. Further investigation is required to understand why they seem to group that way, but the table has provided the impetus.

The highlighting of single entries points out the unusually high rate of transport accidents in Canada and Austria as well as the unusually low rates of death due to natural factors in the U.S. and Canada. The determination that these values are indeed unusual was made by additional calculations in support of the display (subtract out row and column effects and look at what sticks out). But the viewer can appreciate the result without being aware of the calculations. The techniques of spacing tables commensurate with the values of their entries and highlighting unusual values are often useful, but they are not as universally important as the three rules mentioned previously.

The version of table 1 shown as table 5 is about as far as we can go. It may be that for special purposes other modifications might help, but table 5 does allow us to answer readily the four questions about these data phrased earlier. Some aspects are memorable. Who can forget the discovery of the gigantic disparity between the homicide rate in the United States and that of the other fifteen nations reported.

CHAPTER 11

A Rose by Another Name

Eponymy—The practice of affixing the name of the scientist to all or part of what he has found, as with the Copernican system, Hooke's law, Planck's constant, or Halley's comet.

Stigler's law of eponymy—No scientific discovery is named after its original discoverer.*

Nightingale and Her Rose

Florence Nightingale used statistical graphics effectively in her campaign to improve the sanitary conditions of the British army after the Crimean War. She was one of the earliest in Britain to see the power of William Playfair's graphical tools to convey complex statistical information to a broad audience. She drew a series of graphs (which in her correspondence she called *coxcombs* because of their shape and color) that are a variation on Playfair's pie charts. These graphs, now often referred to as Nightingale roses,† are like pie charts in that they use the area of segments of a circle to convey amounts. They are unlike pie charts in two respects. In a pie chart each segment shares a common radius but has a central angle that varies as a function of the data. In a Nightingale rose, each segment subtends the same angle from the center, but it is the square root of the radius that varies with the data. An important second difference is that Nightingale roses can be used to communicate complex information effectively.

One of Nightingale's original roses is shown in figure 1, which (as she said in the facing text) allows us to *"estimate how excessive was the rate of mortality suffered under the conditions of the first winter, and how low it fell under the sanitary conditions of the summer."* Another of her Roses divides the mortalities into battle-related and non-battle-related. This second graph makes it abundantly clear that for the British soldier the least dangerous aspect of the Crimean War was the opposing army.

*Stigler (1980) credits this to Robert Merton (1973).

†Described in 1857 by Nightingale in an anonymous, privately printed and distributed publication entitled *Mortality of the British Army*. This was amplified in her larger report, *Notes on Matters Affecting the Health, Efficiency and Hospital Administration of the British Army* (London, 1858) with many examples.

Florence Nightingale

FIGURE 1. Florence Nightingale's rose, showing the mortalities of the British army during two years of the Crimean War. The dotted circle represents what mortality would have been had the army been as healthy as inhabitants of the city of Manchester (mortality rate of 12.4 per 1,000) during the same time period

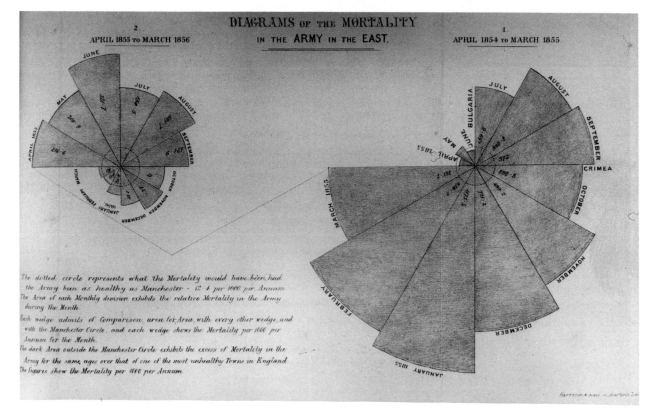

Four Modern Roses

The Nightingale rose has been used in precisely the same way in modern times. Figure 2 is a chart from the *New York Times* purporting to show an annual pattern of incidence of Hodgkin's disease. This application, although using Nightingale's format, does not pack the same punch. This is for two reasons. The first is because the phenomenon being displayed is not as clearly linked to the time of year as was British mortality in the Crimea. Second, the graph is drawn incorrectly, since the radius of each segment varies as the number of Hodgkin's cases, rather than as its square root. Thus our eyes, which see areas, are deceived.

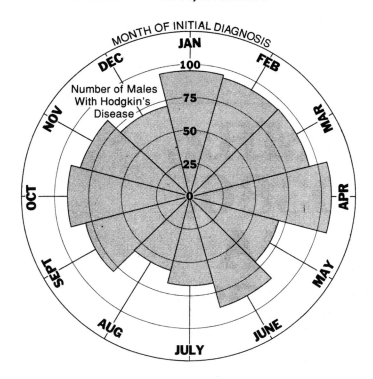

HODGKIN'S DISEASE IN SPRING
Certain cells of the immune system follow a yearly cycle; for example, T cells fall to their lowest level in June. There are some indications that people with Hodgkin's disease are more likely to be diagnosed in the spring and some scientists think the two cycles are linked.

FIGURE 2. A flawed roselike chart from the *New York Times* (October 2, 1990, p. C1) depicting the annual pattern of incidence of Hodgkin's disease.

A strength of the rose is that all segments have the same angle. This means that in a repeated sequence of roses with different data, corresponding segments in different roses are always in the same relative position. This allows us to be able to make effective comparisons. This characteristic makes the rose especially suitable for displaying contingency table data.

Consider the display in figure 3,[1] which shows the standardized percentage distribution of linguistic proficiency among 75,235 Canadian public employees in 1972 broken down by their native language. The instantaneous perception is that Francophones are as bilingual as Anglophones are unilingual. A pie chart (figure 4) of the same data emphasizes the character of the standardization, but diminishes our perception of the two-by-twoness of the data. The capacity of explicitly showing the contingency structure of the data becomes even more of an asset when we look at multiple displays.

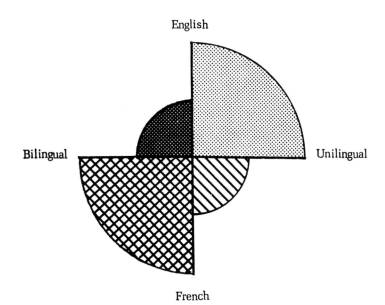

FIGURE 3. A two-by-two rose-like chart prepared by Stephen Fienberg (1975), showing the standardized percentage distribution of linguistic proficiency among 75,235 Canadian public employees in 1972 broken down by their native language

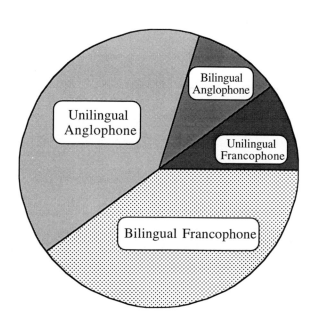

FIGURE 4. A standard pie chart of the same data as in figure 3

Respiratory Symptoms among Coal Miners
All Ages Combined

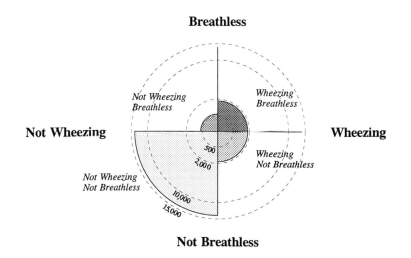

Respiratory Symptoms among Coal Miners for Nine Age Groupings

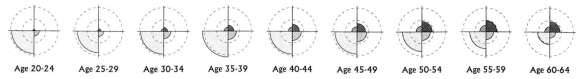

In 1970, Ashford and Sowden[4] provided the data from which figure 5 was constructed. This sequence of nine two-way icons shows a chilling story of how time in a mine affects pulmonary health, although a parallel plot of a matched sample of men who work in a more benign environment would make the obvious causal inference more credible.

The nine component icons in figure 5 are arranged in a row, as befits the linear aspect of the underlying dimension of age. But in other circumstances a different arrangement can be more informative. As we shall see in figure 7, arraying similar icons suitably on a map can make very good sense. But modern social data are commonly in the form of a multidimensional contingency table. Such tables are often complex and difficult to understand. In a 1979 study directed by Jennie McIntyre, 247 women who had been victims of a sexual assault were interviewed. Figure 6 was an attempt by Albert Biderman and Douglas Neal to summarize the results of those interviews. The data reflect the outcome of the attack (shrunk to a two-by-two categorization that is represented as a single icon), the situation in which the attack took place (four possibilities in a nested structure), and four possible modes of resistance. The two-by-two rose that represents

FIGURE 5. Respiratory symptoms among coal miners shown as two-way roses aggregated over all ages and broken down by age. The nine age-specific roses show the progression of pulmonary problems in miners.

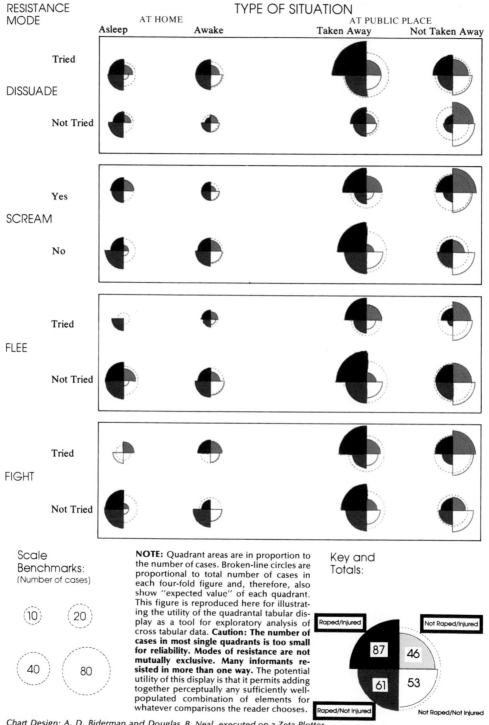

Frequencies of Rapes and Injuries in Various Types of Stranger Rape
Situations by Whether or Not a Mode of Resistance Was Used
Quadrantal Tabular Graph

FIGURE 6. An multivariate application of the two-way rose prepared by Albert
Biderman and Douglas Neal in 1979. (Reprinted from the spring-summer 1979,
vol. XIII, 2–3, newsletter of the Bureau of Social Science Research, Washington,
D.C.) (See insert for color version.)

outcomes is used as a plotting icon within the structure of the entire data set. Despite the complexity of the data, we can see at a glance that the most dangerous situation is being taken away from a public place, and that in any situation the mode of resistance chosen does not seem to have much effect.

Whose Rose?

The euphonious eponymy "Nightingale roses" combined with my knowledge of Stigler's law (quoted earlier) lit a warning light. But if not Nightingale, who? The beginning of an answer is found in Funkhouser's description of the work of the French engineer Léon Lalanne.[3] Lalanne plotted the relative frequency of winds as a function of their direction over the framework of a compass (see figure 7). Lalanne's design is remarkably similar to those Matthew Fontaine Maury of the U.S. Navy, who published from 1847 onwards wind and current charts from logbooks "stored as rubbish" in his office. The winds were identified by the feather end of an arrow, with the character of the feathering showing the character of the wind and the color indicating the season. On his pilot charts Maury used these wind roses to show how the winds blow in each month and in every five-degree square of the ocean.

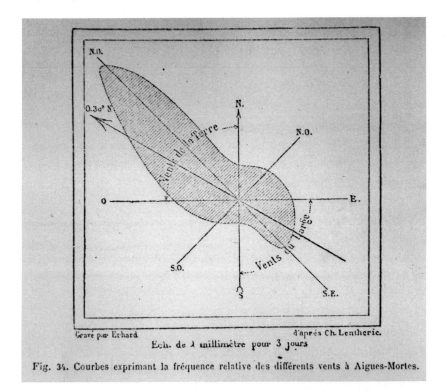

Fig. 34. Courbes exprimant la fréquence relative des différents vents à Aigues-Mortes.

FIGURE 7. A pre-Nightingale rose prepared in 1830 by Léon Lalanne showing the prevalence of winds.

These "wind rose" charts are clearly forebears to Nightingale's coxcombs. They are also obvious aids to navigators. Certainly, if one was trying to sail from here to there, a map with such graphical icons

FIGURE 8. A more literal icon used to show wind direction from Juan de La Cosa's 1500 map of the New World. The map is notable for its depiction of Cuba as an island. In addition, could this "wind gnome" be the eponymous source for the term "gnomon" that is used to describe the gadget that shows the direction of the light on sun dials? (See insert for color version.)

*These names are used by Nils Adolf Erik Nordenskiöld (1832–1901) in his *Facsimile Atlas* (1889). There are primitive wind-roses in Angelino Dulcent's 1339 map of the Baltic and on Petrus Vesconte's 1318 map of the Black Sea.

spotted cleverly around on it might have been invaluable. Had someone used this earlier? Apparently not, for although the name "wind rose" is ancient, those earlier versions were highly stylized and used principally for decoration. Other devices were employed for winds. One example of these (figure 8) is taken from a map drawn in 1500 by Juan de La Cosa, who served as navigator on the *Niña* during Columbus's second voyage (1493–94).

In the search for early users of the sort of Lalanne-Maury wind roses, I stumbled onto a reference to Andrea Bianco, an Italian cartographer living in Venice in the early fifteenth century. He published charts in 1426 that contained many wind roses. (In the nineteenth century these were called more poetically "roses-of-the-winds" or less poetically, but more accurately, "compass cards".)* Bianco's charts are reasonably well known to cartographers, not because of their inclusion of wind roses, but rather because they contained a depiction of islands west of the Azores ("Antillia" and "De la man Satanaxio"). This has been the evidence that has led many to believe that he possessed some knowledge of North and South America.[4]

Looking forward from Nightingale's imaginative work has led us to a graphical format that can surely prove useful in the depiction of complex contingency tables in addition to the kinds of cyclic data that led her to adopt the format. But looking backward to the users upon whose work Nightingale built has brought to us a wonderful tale of discovery.

CHAPTER 12 Trilinear Plots

In his *Atlas* of 1786, William Playfair wrote of the increasing complexity of modern life. He pointed out that when life was simpler and data were less abundant, an understanding of economic structure was both more difficult to formulate and less important for success. But by the end of the eighteenth century, this was no longer true. Statistical offices had been established and had begun to collect a comprehensive spectrum of data on which political and commercial leaders could base their decisions. Yet the complexity of these data precluded their easy access by any but the most diligent.

Playfair's genius was in surmounting this difficulty through his marvelous invention of statistical graphs and charts. In the explanation of his innovation he tells the viewer:

> On inspecting any one of these Charts attentively, a sufficiently distinct impression will be made, to remain unimpaired for a considerable time, and the idea which does remain will be simple and complete, at once including the duration and amount. Men of great rank, or active business, can only pay attention to general outlines; nor is attention to particulars of use, any further than as they give a general information: And it is hoped, that with the assistance of these Charts, such information will be got, without the fatigue and trouble of studying the particulars of which it is composed.[1]

The complexity of life within eighteenth-century Britain and the massiveness of available data are but trifles in comparison to today's complex network of data sources and topics. These data are being transformed into graphic forms at a breathless pace.

This graphic explosion, though caused by the need to present massive amounts of information compactly, is abetted by the computer, which can produce instructions for graphic output as easily as it can crunch numbers. Moreover, the means of disseminating graphics have advanced almost apace with the means of producing them. Today's printing techniques need not distinguish between word and image— the page is merely a matrix of white and black to be arranged. Thus, as the need for graphics has increased, the means for producing and reproducing them have improved.

All the pieces are here—huge amounts of information, a great need to clearly and accurately portray them, and the physical means for doing so. What has been lacking is a broad understanding of how best to do it.

In the recent past I have seen repeatedly a particular data structure in the news media. In all cases they have been displayed in a way that hindered comprehension. The data structure I am referring to is technically termed a three-dimensional probability simplex. Specifically, they are a series of three number sets, each of which sums to one. Such data show up in economics (percentage of each country's economy in agriculture, in manufacturing, and in service), in sports (percentage of each football team's offense due to rushing, passing, and kickoff returns), in politics (percentage of electorate in each state for Dole, Clinton, and Perot), and even in the budgets that we submit for support (percentage of salary, benefits, and overhead).

These data are usually presented as a table or as a sequence of pie charts with three sectors in each pie. With a modest-sized data set a well-designed table can sometimes be helpful. Any help obtained from a set of pies is almost surely an accident.

The trilinear plot is a too-seldom-used display for this particular kind of data. Guilbaud[2] used it successfully almost fifty years ago, and Coleman[3] more recently. I am sure that much earlier as well as more recent examples can be found.

The trilinear plot uses the fact that since the three-dimensional structure of the data is illusory (if you know two of the numbers you can compute the third), you can plot them unambiguously in two-dimensional space. Figure 1 is a skeleton of such a plot, showing its essential characteristics.[4] The components are labeled perpendicularly to their associated axes, which run from each apex to the midpoint of the opposite baseline.

As is always the case, the value of any graphical figuration must be determined from its usage. To help in making this judgment let me offer two applications. The first is my own, plotting data from the National Assessment of Educational Progress (NAEP). The second is a remarkably apt use from Graham Upton's splendid 1994 paper on the 1992 British general election.

Example 1: Comparing States in the National Assessment

Student performance on the tests of the National Assessment of Educational Progress (NAEP) has been characterized through a formal judgmental procedure into four levels: Advanced, Proficient,

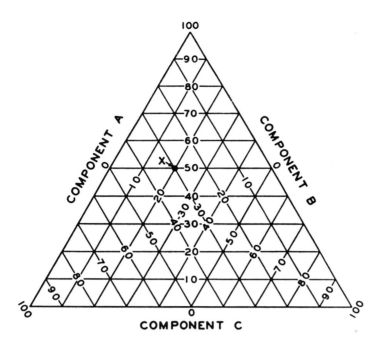

FIGURE 1. An explanatory format for the trilinear plot taken from Schmid and Schmid, 1979, p. 151. (Reproduced with permission.)

Basic, and Below Basic. Although the definitions of exactly what kinds of proficiencies constitute each level vary by age and subject matter, these levels are now in use in the math and verbal assessments. Moreover, current plans anticipate that they will eventually permeate all of the NAEP testing areas.

Because these performance levels are considered important for a variety of policy decisions and as a criterion-based measure of school effectiveness, many tables are produced and printed that report the percentage of children that score in each of these categories for each NAEP jurisdiction. These tables are produced for each test separately for each of many demographic variables (e.g., ethnicity, sex, community type, parental education). A sample table is shown as table 1.

This table is complete and allows the easy extraction of any state's data. But aside from providing the national mean scores, it does not yield any intuitive view of the distribution of performance across all of the states. Such a view is important, for example, in making comparisons among various demographic subgroups. It is toward providing such an effective display that I utilized trilinear plots.

A trilinear plot would be possible if we could somehow reduce the number of achievement levels to three without losing too much information. Sadly for the enterprise of American education, but happily for graphical display, the level "Advanced" is practically empty. From table 1 we can see that in no state are there more than 6 percent of the pupils at this high level, with the national average at three percent. It is clear that for most purposes very little information would be lost if we combine Advanced and Proficient into a single category. After doing this we can immediately use trilinear plots.

TABLE 1

Overall Average Mathematics Proficiency and Achievement Levels

Grade 8 - 1992

PUBLIC SCHOOLS	Average Proficiency	Percentage of Students At or Above Advanced	Percentage of Students At or Above Proficient	Percentage of Students At or Above Basic	Percentage of Students Below Basic
NATION	266 (1.0)	3 (0.5)	23 (1.1)	61 (1.2)	39 (1.2)
Northeast	267 (3.0)	5 (1.4)	25 (3.0)	59 (3.9)	41 (3.9)
Southeast	258 (1.2)	1 (0.4)	16 (1.0)	53 (1.6)	47 (1.6)
Central	273 (2.2)	3 (0.7)	28 (3.0)	70 (2.8)	30 (2.8)
West	267 (2.1)	4 (1.1)	24 (2.1)	62 (2.7)	38 (2.7)
STATES					
Alabama	251 (1.7)	1 (0.3)	12 (1.1)	44 (2.0)	56 (2.0)
Arizona	265 (1.3)	2 (0.4)	19 (1.4)	61 (1.8)	39 (1.8)
Arkansas	255 (1.2)	1 (0.3)	13 (1.0)	50 (1.7)	50 (1.7)
California	260 (1.7)	3 (0.7)	20 (1.4)	55 (2.0)	45 (2.0)
Colorado	272 (1.1)	2 (0.5)	26 (1.3)	69 (1.3)	31 (1.3)
Connecticut	273 (1.1)	4 (0.6)	30 (1.1)	69 (1.4)	31 (1.4)
Delaware	262 (1.0)	3 (0.4)	18 (1.1)	57 (1.2)	43 (1.2)
Dist. Columbia	234 (0.9)	1 (0.2)	6 (1.0)	26 (1.3)	74 (1.3)
Florida	259 (1.5)	2 (0.4)	18 (1.3)	55 (1.9)	45 (1.9)
Georgia	259 (1.2)	1 (0.3)	16 (1.0)	53 (1.5)	47 (1.5)
Hawaii	257 (0.9)	2 (0.4)	16 (0.8)	51 (1.2)	49 (1.2)
Idaho	274 (0.8)	3 (0.4)	27 (1.2)	73 (1.1)	27 (1.1)
Indiana	269 (1.2)	3 (0.4)	24 (1.3)	66 (1.5)	34 (1.5)
Iowa	283 (1.0)	5 (0.7)	37 (1.4)	81 (1.2)	19 (1.2)
Kentucky	261 (1.1)	2 (0.4)	17 (1.1)	57 (1.3)	43 (1.3)
Lousiana	249 (1.7)	1 (0.2)	10 (1.2)	42 (2.0)	58 (2.0)
Maine	278 (1.0)	4 (0.6)	31 (1.9)	77 (1.3)	23 (1.3)
Maryland	264 (1.3)	4 (0.6)	24 (1.3)	59 (1.5)	41 (1.5)
Massachusetts	272 (1.1)	3 (0.5)	28 (1.4)	68 (1.5)	32 (1.5)
Michigan	267 (1.4)	3 (0.5)	23 (1.7)	63 (1.6)	37 (1.6)
Minnesota	282 (1.0)	6 (0.7)	37 (1.2)	79 (1.2)	21 (1.2)
Mississippi	246 (1.2)	0 (0.2)	8 (0.8)	38 (1.5)	62 (1.5)
Missouri	270 (1.2)	3 (0.4)	24 (1.3)	68 (1.6)	32 (1.6)
Nebraska	277 (1.1)	4 (0.5)	32 (1.9)	75 (1.2)	25 (1.2)
New Hampshire	278 (1.0)	3 (0.6)	30 (1.5)	77 (1.0)	23 (1.0)
New Jersey	271 (1.6)	4 (0.6)	28 (1.4)	67 (1.8)	33 (1.8)
New Mexico	259 (0.9)	1 (0.3)	14 (1.0)	54 (1.4)	46 (1.4)
New York	266 (2.1)	4 (0.6)	24 (1.6)	62 (2.3)	38 (2.3)
North Carolina	258 (1.2)	1 (0.3)	15 (1.0)	53 (1.5)	47 (1.5)
North Dakota	283 (1.2)	4 (0.6)	36 (1.7)	82 (1.3)	18 (1.3)
Ohio	267 (1.5)	2 (0.5)	22 (1.4)	64 (2.0)	36 (2.0)
Oklahoma	267 (1.2)	2 (0.3)	21 (1.2)	65 (2.0)	35 (2.0)
Pennsylvania	271 (1.5)	3 (0.7)	26 (1.5)	67 (1.7)	33 (1.7)
Rhode Island	265 (0.7)	2 (0.3)	20 (1.3)	62 (1.2)	38 (1.2)
South Carolina	260 (1.0)	2 (0.5)	18 (1.1)	53 (1.2)	47 (1.2)
Tennessee	258 (1.4)	1 (0.4)	15 (1.2)	53 (1.8)	47 (1.8)
Texas	264 (1.3)	4 (0.6)	21 (1.4)	58 (1.5)	42 (1.5)
Utah	274 (0.7)	3 (0.5)	27 (1.1)	72 (1.3)	28 (1.3)
Virginia	267 (1.2)	3 (0.5)	23 (1.4)	62 (1.6)	38 (1.6)
West Virginia	258 (1.0)	1 (0.2)	13 (0.9)	53 (1.5)	47 (1.5)
Wisconsin	277 (1.5)	4 (0.6)	32 (1.4)	76 (1.9)	24 (1.9)
Wyoming	274 (0.9)	2 (0.5)	26 (1.0)	73 (1.3)	27 (1.3)
TERRITORIES					
Guam	234 (1.0)	1 (0.2)	7 (0.7)	30 (1.4)	70 (1.4)
Virgin Islands	222 (1.1)	0 (0.1)	1 (0.3)	13 (1.0)	87 (1.0)

Figure 2 is a trilinear chart that shows the performance of the entire U.S. as well as that of Iowa and the Virgin Islands, on the 1992 eighth-grade NAEP State Mathematics Assessment. The arrows springing from each jurisdiction intersecting perpendicularly to the three axes show how the points should be interpreted. Thirty-seven percent of Iowa's students performed at the Advanced or Proficient level, compared to only one percent of those from the Virgin Islands and twenty-three percent for the nation as a whole. Forty-four percent of Iowa's students were at the Basic level compared to twelve percent in the Virgin Islands. Last, only nineteen percent of Iowans were Below Basic compared to eighty-seven percent in the Virgin Islands.

Obviously, these are two extreme points, but they serve to illustrate how to read the chart. Intuition is aided by noting the direction and distance one must traverse in moving from the Virgin Islands to Iowa. In geographic terms, moving east is good, moving north is better; in general, any jurisdiction that is northeast of another dominates the latter in both the Basic and Proficient & Advanced categories.

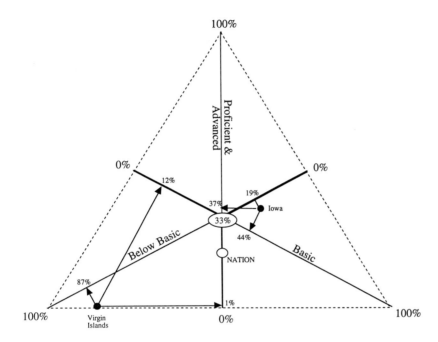

FIGURE 2. An illustration of the trilinear format displaying the performance of two jurisdictions in the 1992 eighth-grade NAEP State Mathematics Assessment.

Showing two points in a comprehensible way is no trick. How well does this display method allow us to look at large quantities of data? Shown in figure 3 are all forty-four of the participating jurisdictions (forty-one states, Guam, the Virgin Islands, and the District of Columbia), as well as an open circle representing the national average. We have simplified the picture somewhat by omitting portions of the axes lines. In so doing we have formed and labeled three "tridants" (surely not quadrants). Any jurisdiction that falls into one of these tridants has that tridant as its modal level. Thus, for example, Iowa, Minnesota, and North Dakota have more eighth graders at the Basic level than at either of the other two, whereas Guam, Mississippi, and Louisiana, have more eighth graders at the Below Basic level than at any other. It is discouraging to note that no jurisdiction's point falls into the Proficient & Advanced sector.

One can also incorporate comparisons by various sorts of demographic variables. Figure 4 shows all of the forty-four jurisdictions broken down by two extreme levels of parental education: children whose parents were college graduates vs. children whose parents did not graduate high school. The points representing New Jersey are joined, as are, for purposes of comparison, the points for the nation as a whole.

FIGURE 3. A trilinear depiction of the performance of all forty-four participating NAEP jurisdictions (twenty-four identified) on the 1992 eighth-grade NAEP State Mathematics Assessment.

FIGURE 4. A trilinear depiction of the performance of all forty-four participating NAEP jurisdictions on the 1992 eighth-grade NAEP State Mathematics Assessment stratified by two extreme levels of parental education. The state of New Jersey and the nation as a whole are explicitly identified.

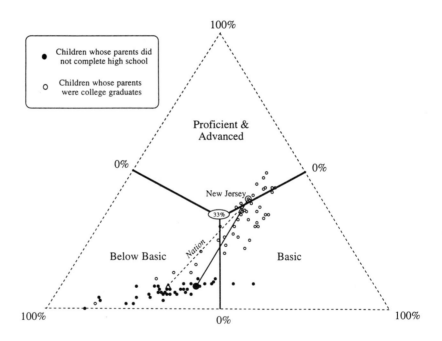

Simplifying the display by removing some of its explanatory elements is a reasonable thing to do once the reader has become accustomed to its character. A graph as full of help as the one pictured in figure 1 leaves little room for the data. One obvious direction for increasing the richness of the data would be an expansion of figure 4 by deleting the points entirely and substituting jagged lines for each jurisdiction that connect the (now invisible) points associated with several levels of parental schooling. How many lines can be visually accommodated is an empirical question. But the tale told by such a single image, were it comprehensible, would be nothing less than a state-by-state depiction of the effect of parents' education on children's mathematics performance. The variability among states shown within a single stratum of parental education helps us to understand better the extent of the difference that schools can make. As such, making this statistical conditioning accessible seems a worthy goal.

Example 2: Political Change in England 1987–92

Graham Upton sought to represent graphically the percentage vote for each of Britain's three political parties in the 1987 and 1992 general election. Each constituency was plotted with a point. The number of constituencies, hence the number of seats in parliament, that fall onto the same place in the plane are represented by the size of the plotted point. In this instance, dividing the trilinear plot into tridants has special importance, for a constituency falling into a particular tridant means that the party represented by that tridant has won the seat for that constituency. The security of that party's hold on the seat is represented by the distance of the point from the tridant's boundaries. The results of the two elections are shown in the two panels of figure 5. The observation that there are many more constituencies near the Conservative-Liberal boundary than near the Labour-Liberal boundary suggests that if you are a Liberal Party strategist, you should run your strong candidates and put your effort in Conservative rather than Labour constituencies.

One can compare the gross features of the two elections from these figures. One can see "the general reduction of the Liberal vote and the shift of the political centre of gravity towards Labour." (Upton) But the interelection changes that are only hinted at in figure 5 can be

FIGURE 5. Constituency variation in England in (a) 1987 and (b) 1992. (Reproduced from Upton, 1994, with permission.)

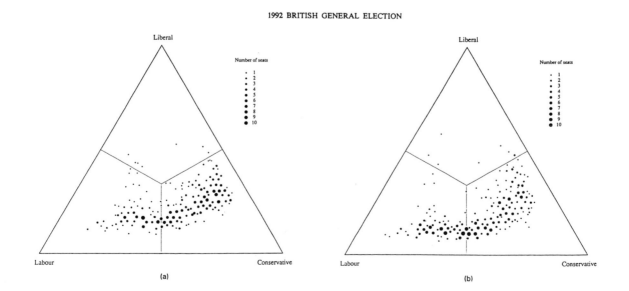

1992 BRITISH GENERAL ELECTION

seen much more explicitly by plotting both elections in the same chart and joining the points representing the same constituency with an arrow pointing toward 1992. This is shown in figure 6.

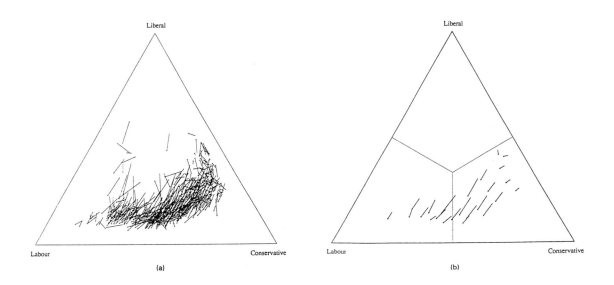

FIGURE 6. (a) Voting changes 1987–92 for individual English constituencies; (b) aggregated voting changes, England 1987–92. (Reproduced from Upton, 1994, with permission.)

The individual arrows are not easy to see in panel (a). In panel (b), which is an aggregation taken over constituencies of similar location in 1987, we see the "political equivalent of a magnetic field, with the flow of the arrows showing remarkable consistency."[5] Most of the large arrows are roughly parallel to the left-hand side of the triangle. This suggests that "the Conservative vote was largely unaltered whereas varying numbers of Liberal supporters shifted allegiance to the Labour party."[6]

Concluding Observations

Despite their frequent suitability, trilinear plots are rarely seen in the media. Why? Certainly part of the reason must be convention. There is a cost involved in bucking convention and using an innovative graphical form. Thus, whenever an innovative graphical format is proposed, an important consideration must be the gains associated with the new form versus the losses associated with moving away from the conventional display. Pie charts are used despite their flaws because they are a conventional and obvious metaphor. Trilinear plots are not in common use, and despite their obvious appropriateness in these applications, they take some getting used to. It was my intention with the two examples and their variations shown here to provide the reader with some experience, and hence comfort, with the format. By my doing so, perhaps others will produce evocative applications of this somewhat specialized format. Thus can we expand the public consciousness of our graphical repertoire and continue to increase the comprehensibility of information.

CHAPTER 13 Implicit Graphs

I suspect that not all readers of chapters 6 and 7 will come away loving
graphical train schedules. This pessimism is at least partially empirical
in origin, for not too long ago I received a critical letter from Stephan
Michelson, of Takoma Park, Maryland, telling me precisely that. His
criticism was clearly thought through and broadly based. Its essence
was that too much space was utilized telling him what he didn't want
to know, leaving little for telling him what he needed to know. He then
suggested that the "Amtrak train schedules [were] just fine."

He was right and wrong. Everything depends on the purpose of
the display. Marey's train schedule provides a wonderful picture of the
pattern of train service between Paris and Lyons. The density of lines
is a fine metaphor for the frequency of trains; their slope for the speed
and direction. But Marey's graph doesn't provide detailed information
about how long a particular train stops in any city, or even whether it
will stop at all. For the purpose of extracting small details about a specific train, the common tabular train schedule may indeed be "just
fine," or at least serviceable. Although if I have a very specific question like What time does the train that leaves New York at 6:17 arrive
in Princeton? a better display still would be "7:35" in 36 point type.
The obvious problem with this sort of single-point, single-purpose
display is that while most people have a single question, not everyone
has the same single question.

How can we fashion displays that are tailored to the user? In the
modern world of high-powered personal computing, a lot of high-
tech answers suggest themselves. But let us explore some lower-tech
possibilities. A common feature at many of the exchange stations of
the Paris Metro is a system map with a large matrix of buttons at the
bottom associated with all possible destinations. Pressing one button
lights up a path from here to there. This display satisfies both goals,
general and specific. Similar displays are also found in museums and
shopping malls. Are there other alternatives?

In Jacques Bertin's encyclopedic *Semiology*[1] is a busy graph (see figure 1) showing the number of people in 1954 in each of France's ninety *départements* (and Paris) broken down further by the three sectors of the work force (I. Agriculture, II. Industry, III. Service). The *départements*, identified by their number, are ordered by the size of the work force within each sector. The lines link the same *département* from one column to another. The closer the order of any two columns, the less numerous are the intersections. Despite this particular example being beyond what Bertin recommends as a suitable length for such a display, one can immediately see the big picture. *Départements* that are heavily agricultural have fewer workers in the other two sectors; but the orderings within the industrial and service sectors seem very similar.

FIGURE 1. A comparison of the three sectors of the work force in each of France's ninety *départements* in 1954. Sector I is Agriculture, Sector II is Industry, and Sector III is Service. (From Bertin, 1973.)

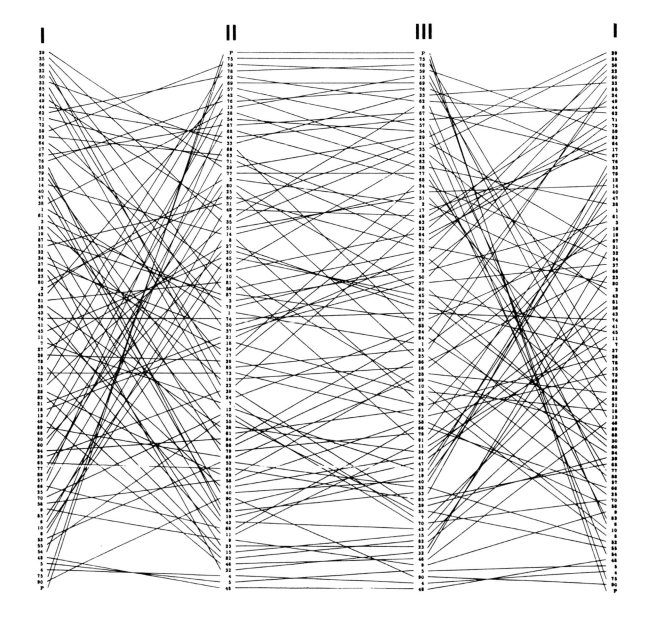

But what about some specific *département?* Finding a particular one and then tracing out its line is difficult and tedious, but not impossible. Suppose we were interested in Hautes Alpes (*département* 5). After carefully searching down each column, we might draw a circle around each 5 we find and then connect the dots with a darker line. In this instance we would discover that it is a small *département* relegated to the bottom of all of the rankings. In fact, if we knew that the goal was always to examine only a single *département,* we might lighten (or even omit entirely) all of the connecting lines and merely reproduce the three ordered columns of *département* numbers. Then each user of the plot would merely have to locate the *département* of choice and connect the dots. The lines, which constitute most of the graph, would be implicit.

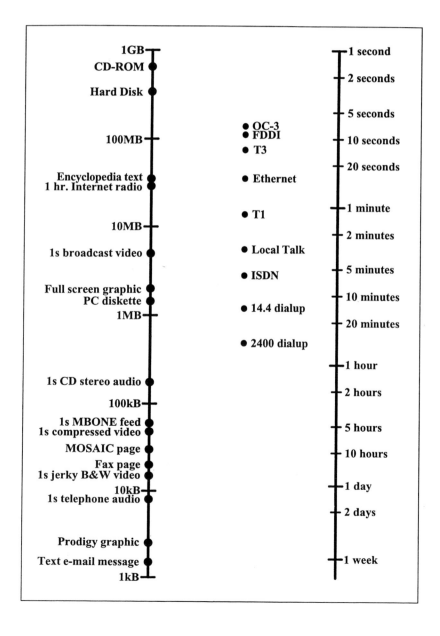

FIGURE 2. An "Internet Trip Time Planner" that allows one to choose the size of the information packet to be sent from the first column, pair it with a mode of transfer from the second, and read off the shipping time from the third. Moreover, one can also choose any permutation of this, by linearly joining any two and reading off the third.

Implicit Display 1: Shipping Times for Information

Shown in figure 2 is a paper version of a slide rule (from *Computers in Physics* 8 (2), March/April 1994, p. 147). The left column (in a log scale ranging from 1kB to 1GB) are some samples of information packets of various sizes that one might want to send electronically. On the rightmost column (also in a log scale) are the times it might take for transmission, ranging from a second to a week. In between these two columns is a third column, representing various modes (speeds) of transmission. To use this graph one merely connects the dots. An e-mail message over a 2400 baud line takes less than ten seconds, a fax page at 2400 baud just under two minutes. We can also use the graph to decide on carriers. For example, if we are to be backing up hard disks often over a network, it tells us we ought not consider anything slower than Ethernet.

This graph is quite satisfactory being left implicit, since the general story is obvious. Big things take longer to send; faster carriers send things more quickly. The important usage is for the specifics.

Implicit Display 2: Constant-Dollar Graph Paper

Figure 3 contains a remarkable implicit graph developed by Robert Sherman. On the vertical axis are dollars in a log scale. This allows us to plot a broad range of prices on the same paper (a television set, a computer, and our monthly pay). Along the horizontal axis is time. Not any time, but specific years—in this case from 1969 to 1982. The grid lines are not horizontal; they are empirical, based on the U.S. Urban Workers Consumer Price Index. All of the graph lines are left implicit. If we plot our salary we need only observe whether it rises or falls relative to the horizontal axis to understand how well we are doing with respect to inflation. Plotting college tuition as well helps us to understand the relationship between the tuition and salary.

Graphs are often thought of as fixed conveyors of specific information. Indeed this is often true. Yet sometimes the information they carry may be implicit, and it is only through an active interaction between the graph and a specific reader that its real functionality emerges. It seems certain that electronic versions of nomographs (an older name for what I have called implicit graphs) are making a comeback. Fixed displays are often a compromise. It seems far more sensible for many purposes to have a data base, and a graphical tool with which to interrogate it. This not only allows the customized preparation of answers, but it also avoids cluttering a plot with material that is not of immediate interest. I have found that being able to flash extra material on and off often yields a display of considerably increased usefulness. Imagine in figure 1 we begin with the full display and then we merely issue a command to "turn off everything that

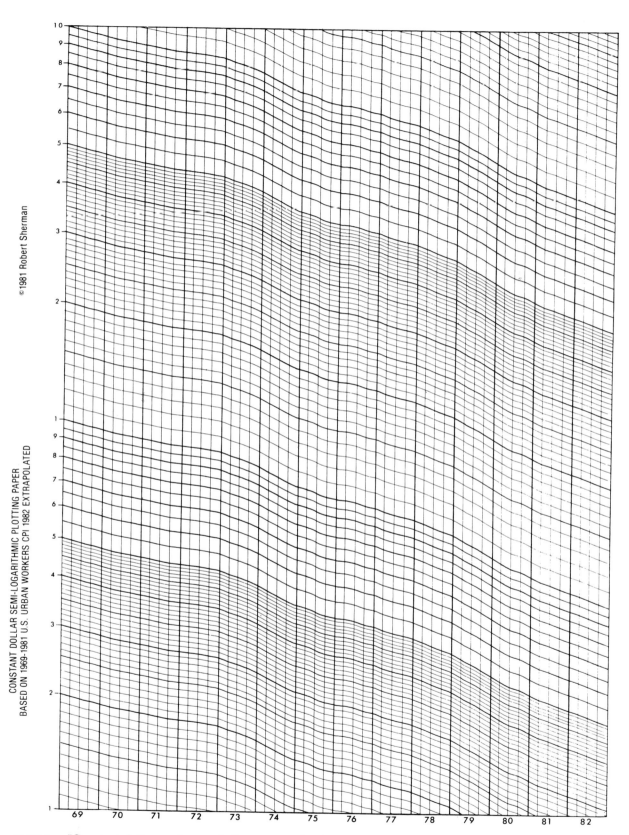

FIGURE 3. "Constant-dollar graph paper" allows one to plot the current money value of anything over the time periods shown and visually compare its change in constant-dollar terms with itself or with other things plotted.

isn't Hautes Alpes." By toggling between the two we can see both the general and the specific. Morcover, someone else who is interested in Val de Marne can use this "graph" equally well.

Language has historically been linear. We tell a story from the beginning to the end. Sure, we can jump around, but all readers of the story would typically jump around in the way that the author plans (although there are surely some who peek at the end to find out how it will turn out, most readers follow the rules). Graphs are nonlinear stories. Every viewer can construct a different tale, although the display's design usually encourages the extraction of some stories more than others. Only recently have such electronic manipulations as hypertext begun to provide language with the same flexibility that graphs have always enjoyed.

It appears that in either word or image, the future of information retrieval is not linear.

Using Graphical Methods

W e frequently encounter data-based graphics in everyday life. If we cannot understand them, our lives are the poorer. In this section are three chapters commenting on different uses of graphics. In chapter 14 a statistical graphic from a report from the U.S. Department of Energy was used as a test question for high-school seniors in the National Assessment of Educational Progress. It was found that half the students surveyed couldn't answer the question correctly. But was it the students who lacked basic "graphicacy" or the constructors of the flawed display?

But the graphs used by Ross Perot in his 1992 presidential campaign weren't flawed. They were clear and informative. In chapter 15 we ponder why graphical summaries of the sort Perot used to communicate the structure of the U.S. budget deficit so effectively were not used by the other candidates.

Data graphics often achieve their power by allowing an immediate visual answer to the question, Compared to what? No phenomenon exists in a vacuum. By providing context a graph facilitates deep understanding. In chapter 16 are three examples of graphics that foster deeper understanding by placing things of little apparent similarity into close physical proximity. Any tool that helps us find "similitude in things to common view unlike" is also likely to help us turn old facts into new knowledge.

The goal of this short section is to illustrate how the proper use of graphical methods provides access to information that aids our understanding of complex but important contemporary issues. It also illustrates how even "experts" in graphical communication screw up seriously upon occasion.

CHAPTER 14

Measuring Graphicacy

A June 1995 newspaper report blared, "Only 50% of American 17 year olds can identify information in a graph of energy sources." Alas!

The graphical item referred to is shown in figure 1. The 50% comprehension was reported at the beginning of June in *From School to Work*.[1] This result was taken *in toto* from one form of the *National Assessment of Educational Progress (NAEP)*, which in turn had taken it from the *Annual Energy Review*. It is a flawed graph in a variety of ways. As it is redrawn (see figure 2) the answer to the question asked is obvious.

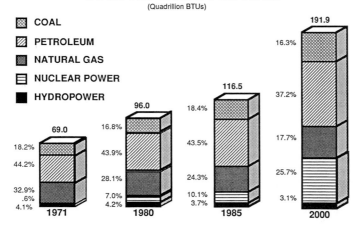

ESTIMATED U.S. POWER CONSUMPTION BY SOURCE
(Quadrillion BTUs)

COAL
PETROLEUM
NATURAL GAS
NUCLEAR POWER
HYDROPOWER

In the year 2000, which energy source is predicted to supply less power than coal?

A Petroleum **D** Hydropower
B Natural Gas **E** I don't know
C Nuclear Power

Source: U.S. Department of Interior   United States Energy Through the Year 2000
BTU: Quantity of heat required to raise temperature of one pound of water one degree Fahrenheit

COPYRIGHT 1973 CONGRESSIONAL QUARTERLY, INC.

FIGURE 1. A graphical test item from the National Assessment of Educational Progress. Its original source was the U.S. Department of the Interior, where it was titled "United States Energy Through the Year 2000." (© 1973 Congressional Quarterly, Inc.)

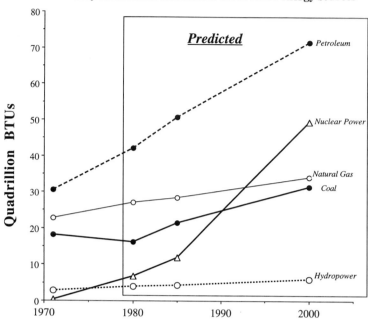

**Profound increases are predicted in the use of Petroleum and Nuclear energy
Only modest increases in the use of other energy sources**

FIGURE 2. One possible revision of the graph in figure 1 that clarifies many things.

*I have often wondered whether the item was considered to have been answered incorrectly if the student responded "E."

Basing a characterization of an examinee's ability to understand graphical displays on a question paired with a flawed display is akin to characterizing someone's ability to read by asking questions about a passage full of spelling and grammatical errors.* What are we really testing?

One might say that we are examining whether or not someone can understand what is de facto "out there." I have some sympathy with this view, but what is the relationship between the ability to understand illiterate vs. proper prose? If we measure the former, do we know anything more about the latter? Yet how often do we encounter well-drawn graphs in the everyday world? Should we be testing what is? Or what should be?

A more practical problem is that if a graph is properly drawn, most commonly asked questions are easily answered. That is the nature of graphics and human information-processing ability. It is harder to ask nontrivial questions of a well-drawn graph. This is not an isolated issue. In the testing of verbal reasoning it is common practice to make a reasoning question more difficult simply by using more arcane vocabulary. This practice stems from the unalterable fact that it is almost impossible to write questions that are more difficult than the questioner is able. When we try to test the upper reaches of reasoning ability, we must find item writers who are more clever still.

While we cannot hope to resolve these issues here, I would like to add one vote toward testing literacy with prose that is correctly composed and testing graphicacy with data displays that adhere to accepted standards of good practice. If we do otherwise we may be able to connect our test with common practice, but is that what we wish to know?

CHAPTER 15 Graphs in the Presidential Campaign: Why Weren't They Used More Broadly?

The 1992 presidential campaign was a heady time for those of us who advocate the increased use of statistical graphics to communicate the complexities of quantitative phenomena. My current hero is Ross Perot. He showed how a complex story can be told through the use of statistical graphics and how such a story can hold an audience of millions spellbound for half an hour. Did he succeed? He was sufficiently convincing to get almost twenty million Americans to vote for him. Why didn't either of the other two candidates use graphs?

Let us begin with an examination of why Mr. Perot's graphs were as successful as they were and then move on to why they were not adopted by the other candidates. It is certainly not because of the graphs' beauty nor their technical virtuosity. Many of Mr. Perot's graphs were often less than wonderful. Many contained very little information and could have been replaced easily by a sentence or a small table.*

*Edward Tufte told me once of a rule of thumb he uses to decide on display format: three numbers or fewer use a sentence; four to twenty numbers use a table; more than twenty numbers use a graph. I may quibble about the boundaries, but I certainly agree with the underlying point.

FIGURE 1. A graph generated by Ross Perot with impact but very little information (Perot, 1992, p. 14).

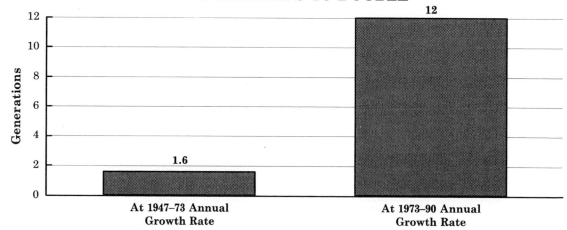

GENERATIONS REQUIRED FOR U.S. LIVING STANDARDS TO DOUBLE

For example, figure 1 might be replaced by the sentence,

> Economic growth over the past seventeen years slowed, so that it now takes almost eight times longer for living standards to double as was the case in the previous twenty-five years.

Or perhaps,

> Between 1947 and 1973 one could expect to live forty-five percent better than one's parents. Since 1973 that figure has shrunk to only six percent better, and it's headed down.

is a little better. In this instance, I prefer the prose to the graph since it states the point explicitly .

Some of Perot's displays, especially his tables, were poorly suited to the broadcast medium because they were too detailed to be resolved successfully on a television screen. Nevertheless, they contained data. As such they stood out starkly against the overstretched rhetoric that was encountered in other political messages. Moreover, these data were then woven into a coherent story of vital concern to Mr. Perot's audience.

As an illustration, let us consider two of Perot's figures in the context of this story. These figures are taken from his campaign book *United We Stand*. The versions of the figures used on television were very similar, although they also contained noninformative color for decoration.

The story begins with the national debt growing exponentially under recent administrations. His evocative plot of the debt (see figure 2) shows both its staggering size as well as its rapid rate of increase. This plot was well done, and although we can make some superficial improvements (correct the overrounded scale on the vertical axis and indicate the various presidential administrations who were in charge

FIGURE 2. An evocative bar chart presented by Ross Perot showing the exponential growth of the national debt over the last twenty-two years (Perot, 1992, p. 7).

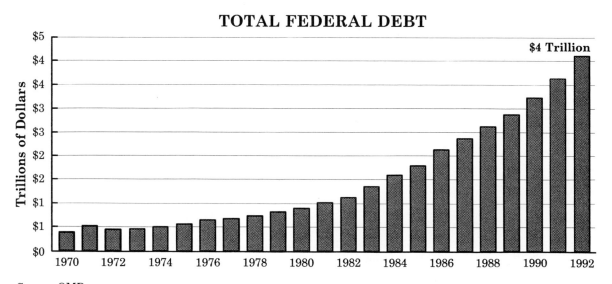

Source: OMB

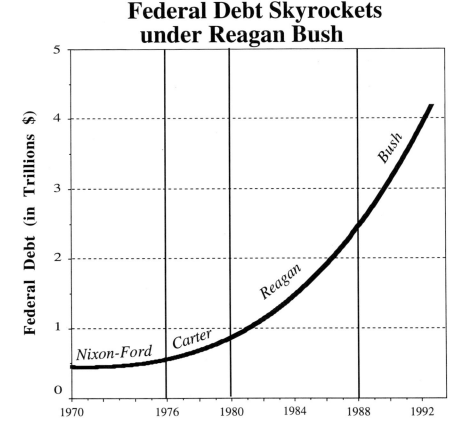

Federal Debt Skyrockets under Reagan Bush

FIGURE 3. A simpler form of figure 2 that also tells explicitly what presidential administration was in office during the years depicted. This version emphasizes the quadrupling of the national debt during the twelve years of the Reagan-Bush administration.

during this time (see figure 3), the basic perceptual story remains the same. I suspect that if one were trying to be less polemical and communicate the story more accurately, the debt could have been shown in constant dollars or as a percentage of GNP.* But aside from making the small rise during Carter's administration disappear, such modifications would not have affected the qualitative picture very much.

With the notion of a staggering debt in place, he explained the implications that its continuance would have and began to discuss how one might slow its increase. The solution was clear; stop spending more then is taken in. How? Perot chose the obvious metaphor to show how the American pie of federal expenditures is divided. Ordinarily, I am unalterably opposed to the use of pie charts, since their message can always be more accurately conveyed in other ways. But here I was almost convinced. Almost. In fact, Perot used two pie charts and a bar chart to tell an important tale (figures 4 and 5).

I have redone his figures as bar charts and linked them. The first one shows that federal entitlements dominate the 1992 federal budget. It illustrates very clearly that the peace dividend, drawing money from the defense budget, has obvious limitations. The obvious implication is that if we must trim $400B from the budget to bring it into

*Mr. Perot knows how to make such adjustments when they suit his purposes. He took this tack when he discussed the growth of federal entitlement programs (Perot, 1992, p. 46). He showed a bar chart in which the first bar was the federal entitlements in 1960 ($26B), the second was the same figure in 1991 dollars ($114B), and the third also adjusted for population growth ($159B). He then adjusted to show the same figure as a function of GNP growth ($286B) and finally topped things off with the actual 1991 figure ($651B).

1992 U.S. GOVERNMENT EXPENDITURES: $1.5 TRILLION

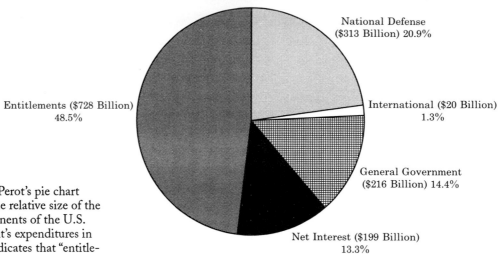

FIGURE 4. Perot's pie chart showing the relative size of the five components of the U.S. government's expenditures in 1992. It indicates that "entitlements" represents the biggest slice (Perot, 1992, p. 45).

1992 FEDERAL ENTITLEMENT OUTLAYS

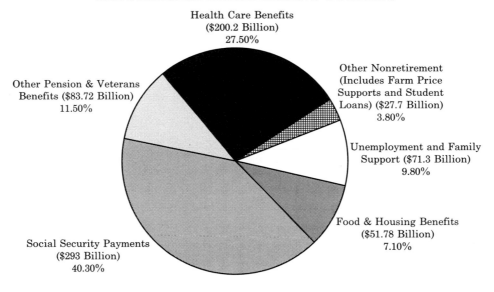

FIGURE 5. A second pie chart dissects entitlements and shows that two-thirds of entitlements are consumed by social security and health care benefits (Perot, 1992, p. 55).

line with the $1.1 trillion in income, there is no other place but entitlements for it to come. This graph screams for the next one, showing the components of the various entitlement programs (figure 6).

A glance at this chart shows why cracking down on welfare cheats and tobacco price supports will not put a dent in the problem. Social Security and Medicare must be the first place to look. The story is inexorable, but even the very gentle suggestion that Mr. Perot made that we cut a little into these entitlements for wealthy retirees was met with storms of protest from representatives of the aged.

The question of why his displays were effective has been answered. They wove an evocative tale, simultaneously providing information

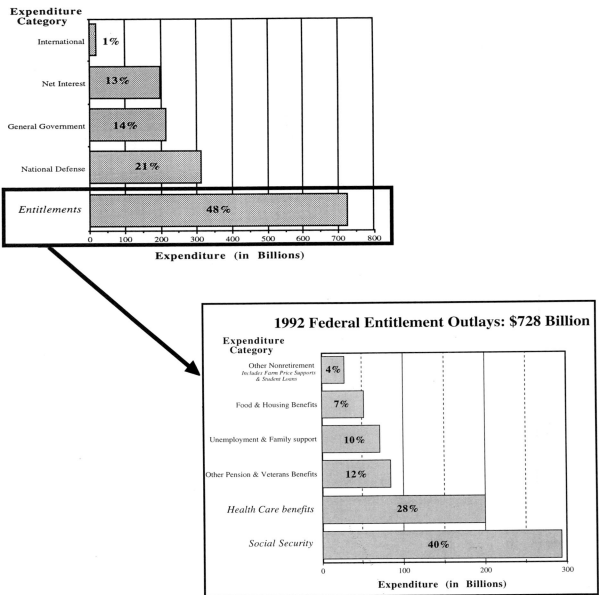

and credibility. But why were graphs not used by the other candidates? Ironically, for the same reason. Even rudimentary graphs tell the story too clearly. No successful candidate could simultaneously do that and still talk credibly about balancing the budget without offending a portion of the electorate whose votes were critical for success.

FIGURE 6. An even more evocative display of the same data is obtained by transforming the pies to bar charts and linking them.

CHAPTER 16 Visual Aids When Comparing
an Apple to the Stars*

*The title for this chapter as well as some of the ideas expressed here are taken from D'Arcy Wentworth Thompson's almost magical *On Growth and Form.* All unidentified quotes come from this source, as well as a variety of paraphrases that are not direct quotes and virtually everything in Latin.

Science, following Chamberlain, is the holding of multiple working hypotheses. A visual depiction allows us to juxtapose a multiplicity of possibilities simultaneously in space. Because this is usually much more efficacious than stringing them together sequentially, as must be done in an aural presentation, this capacity melds well with one of science's principal goals, the search for causes. Even though establishing cause empirically is a task of insuperable difficulty, it merges with another great Aristotelian theme—the search for relations between things apparently disconnected. It has often been found that when we find "similitude in things to common view unlike," we can turn old facts into new knowledge. "Newton did not show the cause of the apple falling, but he showed a similitude between the apple and the stars." Newton was content to be able to bring many diverse phenomena under a few principles, even though the cause of these principles remained undiscovered. Both Hume and Mill declared that all reasoning whatsoever depends on resemblance or analogy, and the power to recognize it.

Jacques Bertin, whose *Semiologie Graphique* has helped many of us to understand better the character of visual communication, pointed out that easing the task of making comparisons is perhaps the highest contribution of graphical display. A number's meaning is enhanced enormously by context. "Our town has five doctors." Is this many or few? Knowing that other towns of the same size average twenty physicians yields a different conclusion than knowing that they average just one. Facts are incomplete without context. Edward Tufte points out (as reported without attribution in chapter 4) that one aspect of Minard's depiction of Napoleon's ill-fated Russian campaign that makes it so poignant is the placement of the river representing the 422,000 men in the attacking Armée de la République adjacent to the trickle of 10,000 men left to retreat. The ready answer this provides to the question "compared to what?" gives it power.

Whether providing context, making far-fetched comparisons, or facilitating unexpected analogies, nothing else approaches the unconscious ease with which a visual display can accomplish these ends. This claim is so self-evident that it hardly seems worthwhile to illustrate it. But once the decision to provide examples is made, the variety of suitable plots is so large as to make the choice very difficult indeed. I have picked two whose topics seem of contemporary interest.

How Crowded Was It?

Among the many things that ran through my mind as I emerged from watching the film *Schindler's List* were images of the crowding depicted: three or four people seemed to share every concentration camp bunk, railroad cars were stuffed to overflowing, apartments in the Cracow ghetto held three or four families to a room. Published sources available to me revealed that these depictions were accurate; each prisoner at the Bergen-Belsen concentration camp barracks was allotted three square feet of space. How crowded was it? Three square feet does not seem like a lot, but can we gain insight into the question by juxtaposing the amount of space in a concentration camp with that in other situations? A graphical answer to this was provided by Albert Biderman and his colleagues in a 1963 technical report to the Office of Civil Defense:

> When we speak about how hot or cold an environment is, the specification of a numerical temperature reading provides immediate meaning references for the reader—his own subjective experience with overly hot and cold environments, extrapolations from his actual experience, knowledge from studies of human physiology, and so on. While square-feet-per-person . . . provide ready physical scales of the intensity of crowding, most of us must undertake a mental exercise to relate a given figure of density to a meaningful experiential context. As an aid to appreciating the significance of measures of crowding stated in terms of physical density, a number of situations of common experience and knowledge have been selected as benchmarks to illuminate scales of . . . density. . . . Various examples of historical instances of overcrowding that have been reviewed have been placed on the same scale.[1]

My rendition of some of the data they reported is shown as figure 1.

To some extent, our understanding is helped by the context. None of us have had first-hand experience in a slave ship or traveling steerage in an nineteenth-century steamship or living in the Black Hole of Calcutta, but being confined to a rush-hour-loaded New York subway car for a year or more is an event that is within reach of our imaginations. But it also shows us that more is needed for the whole story. As can be seen from the graph, there are many common situations that are

How Much Room Was There in ...?

Crowding Situation

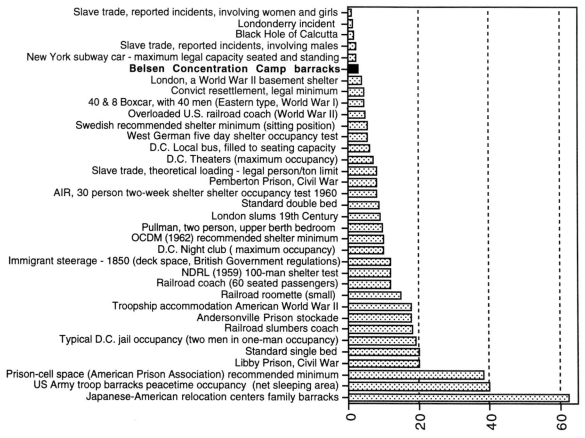

Number of square feet per person

FIGURE 1. A horizontal bar chart depicting the extent of crowding in thirty-five very different situations

quite tolerable and a few that are ordinarily pleasant in which crowding densities are considerably greater than in other situations in which overcrowding has been the source of catastrophe. As is now obvious, to fully understand matters we need information on such variables as (1) time in the space; (2) structural characteristics affecting air supply, heat dissipation, waste disposal, comfort and discomfort; and (3) circumstances that brought people into the crowded situation in the first place.

Even though none of the historical incidents here involves a "pure" case of hardship from crowding alone, this graphic leads us to a fuller understanding both by aiding our intuition and by emphasizing the need for more information.

How Are U.S. Students Doing?

In 1991, U.S. newspapers carried the story of the sorry performance of U.S. 13-year-olds in the International Assessment of Educational

Progress.[2] A graphical rendition of the rankings is shown in figure 2. We see that although the average U.S. student finished far ahead of the average Jordanian, he or she was not ahead of anyone else. The media carried this message loud and clear into all corners of our nation. But what does it mean?

International 1991 Mathematics Assessment
(predicted proficiency for 13-year-olds)

285	Taiwan
284	
283	Korea
282	
281	
280	
279	Soviet Union, Switzerland
278	
277	Hungary
276	
275	
274	
273	France
272	Italy, Israel
271	
270	Canada
269	Ireland, Scotland
268	
267	
266	Slovenia
265	
264	
263	Spain
262	**United States**
.	
.	
.	
.	
246	Jordan

FIGURE 2. A stem-and-leaf diagram showing the place of the United States among the fifteen nations participating in the 1991 International Mathematics Assessment for 13-year-olds.

Placing the results of this assessment into a broader context that includes some characterization of the diversity in our country helps. Figure 3 augments figure 2 by including the mean scores for all forty-one of the participating states.[3] We see that the summarization of the United States with a single number is misleading. In this instance our understanding of either side of figure 3 is aided by the other. Understanding of the scale used here is helped further by noting that the average gain in mathematics proficiency in the United States from fourth grade to eighth grade is forty-eight points. Thus one can think of twelve points as the average gain made in one year. The thirty-nine-point differential between Jordan and Taiwan thus represents about a three-year difference in performance.

FIGURE 3. Back-to-back stem-and-leaf diagrams that expand figure 2 to include the performance of the forty-one states that participated in the 1992 State Mathematics Assessment.

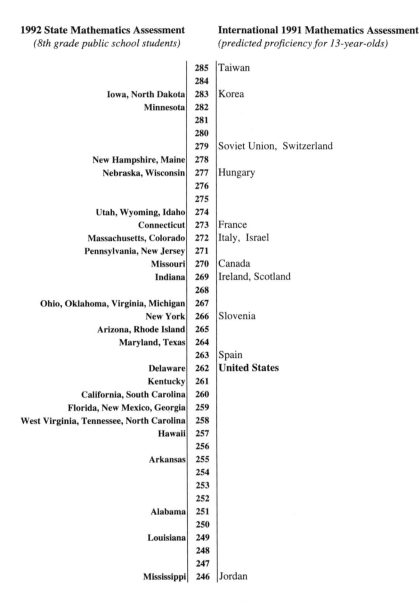

1992 State Mathematics Assessment *(8th grade public school students)*		International 1991 Mathematics Assessment *(predicted proficiency for 13-year-olds)*
	285	Taiwan
	284	
Iowa, North Dakota	283	Korea
Minnesota	282	
	281	
	280	
	279	Soviet Union, Switzerland
New Hampshire, Maine	278	
Nebraska, Wisconsin	277	Hungary
	276	
	275	
Utah, Wyoming, Idaho	274	
Connecticut	273	France
Massachusetts, Colorado	272	Italy, Israel
Pennsylvania, New Jersey	271	
Missouri	270	Canada
Indiana	269	Ireland, Scotland
	268	
Ohio, Oklahoma, Virginia, Michigan	267	
New York	266	Slovenia
Arizona, Rhode Island	265	
Maryland, Texas	264	
	263	Spain
Delaware	262	**United States**
Kentucky	261	
California, South Carolina	260	
Florida, New Mexico, Georgia	259	
West Virginia, Tennessee, North Carolina	258	
Hawaii	257	
	256	
Arkansas	255	
	254	
	253	
	252	
Alabama	251	
	250	
Louisiana	249	
	248	
	247	
Mississippi	246	Jordan

Has finding "similitude in things to common view unlike" helped us, in this instance, to turn old facts into new knowledge? In trying to understand the likely causes of the poor math performance of Mississippi's eighth graders, would we be helped more through comparisons with Jordan than we would with Iowa? Can we not gain some useful insights into what it takes to get good performance by examining the similitudes among Iowa, North Dakota, Korea, and Taiwan? Without further evidence, we cannot tell for sure, but these surely seem, in prospect, like more fruitful paths to pursue than wringing our hands and looking for similarities between the United States and Spain.

How likely is the prospect of successfully understanding the causal variables in education from such data? Is it substantially less than what we might have expected to understand about celestial motion from what we can learn from a graph of an apple falling? (*"Plurimum amo analogias, fidelissimos meos magistros, omnium Naturae arcanorum conscios."*)[4] There is ample evidence to indicate that our purposes will be served sufficiently if through the correlation and equating of diverse phenomena, we can weave a web of connection and interdependence. It will then be a matter of small import if "the metaphysician withholds from that interdependence the title of causality."

SECTION V

Improving Graphical Presentations

This last section is an oddball potpourri of topics, each touching on a smallish aspect of communication that for one reason or another caused me an intellectual itch.

Chapter 17's genesis was the largely anachronistic segregation of text and figures that is practiced by many publishers. In darker moments, I suspect that leaving any apparently unproductive white space on a page must irritate compositors (and perhaps accountants). This inference is drawn from the observation of how often words and images are packed together to optimize space even when it appears to do so at the expense of effective communication.

The principal message in chapter 18 is to not take any set of rules too seriously. The face of beauty has many aspects. But nothing said within this chapter should be interpreted as support for the ubiquitous pseudo-three-dimensional multicolored pie charts favored in corporate annual reports; they remain pariahs in any world that honors grace, elegance, or truth.

Sense-lining, defined and discussed in chapter 19, is a small but fascinating idea. Why be bound by the margins of the page?

Instead, why not stop writing when your idea runs out?

Such a practice provides an important aid to comprehension for the reader, and forces the author to be sure each line has meaning.

Thus despite sense-lining's liturgical origins, it can also be of secular value.

Chapter 20 is the result of some truly rotten talks I have sat through in what seems like an unbroken sequence of long-forgotten conferences. I have found that speakers at conferences are often very enthusiastic about using acetate slides and an overhead projector to show an enormous number of unbelievably uninteresting details to an often soporific audience. The overhead slides are usually prepared by photocopying single-spaced typescript onto the acetate. The results are unreadable any farther away from the screen than the presenter. The audience's choice is either to doze off or to fantasize in the absence of facts. I have done both, but sometimes, since I can't see the content of the presentations, I think about their form. On one undistinguished occasion, I found my old friend Jim Ramsay suffering beside me. Afterwards we retired to a local watering hole, and with the able assistance of Molson Export, we put together some rules for making readable overhead slides. Chapter 20 contains those rules, as well as some obiter dicta about effective presentations.

CHAPTER 17 Integrating Figures and Text

It's June 17 and ten o'clock in the morning. The first mail delivery has arrived and with it a promising envelope from Byrd Press, of Richmond, Virginia. I tear it open and find, much to my delight, the latest issue of *Chance* magazine. The cover has a picture of Wayne Gretzky, and the table of contents reveals a variety of articles of great potential. I can't wait to read them. But as always, before anything else, I turn quickly to the latest from my favorite columnist. "Oh no," I moan, "they screwed up the compositing." My article (chapter 9 in this book) on double Y-axis graphs has been twisted to fit onto two pages by shrinking the figures so much that their details are unreadably small and they are all squished onto a single page facing a single page of text (figure 1).

FIGURE 1. Pages 50 and 51 of *Chance*, 4(1), 1991—one compositor's idea of an effective way to array figures and text

VISUAL REVELATIONS

Howard Wainer, Editor

Double Y-Axis Graphs

The use of a pie chart is sinful, but it is venial. Using "Double Y-Axis Graph" is mortal. If there is a just God I am sure that there is a special place in the Inferno reserved for its perpetrators.

A *double y-axis graph* is a format that refers one function to the left axis and a second function to the right. Most graphics packages offer this type of graph as an easily used option. It allows one to readily twist the facts to suit your aims.

Figure 1 provides a rough duplicate of a plot that appeared in the 1964 Surgeon General's report *Smoking and Health*. The legend has been modified to be more informative. One could, puckishly, consider what the reaction to such a graph might be from defenders of the tobacco industry. I imagine that a statistician working in this industry who brought such a figure to his boss, as an example of clear data display, might be asked to rework the plot. The double y-axis format in Fig. 2 is just the thing for obscuring these data. In Fig. 2 is such a plot (worthy of the name) with a suitably informative legend.

Of course, this transformation of an informative graph into one that misinforms is, to the best of my knowledge, hypothetical. Is this potentially confusing format actually used? You betcha!

Figure 3 is taken from the May 14, 1990 issue of *Forbes* magazine. It purports to show that while per pupil expenditures for education have gone up precipitously over the last decade, student performance (as measured by mean SAT scores) has not responded. The conclusion, of course, is that we ought not waste our money on education.

Of course, such inferences based on these data are completely specious. Since we have complete control of both Y-axes we can rescale to yield any result we want. Consider the alternatives shown in Figs. 4 and 5.

In Fig. 4 we scale both Y-axes to cover the range of the data (just as was done in Fig. 2). This yields the conclusion that both variables move apace. In Fig. 5 we stretch the expenditure axis from $0 to $20,000, thus making the educational outlays appear penurious. At the same time we expand the SAT axis so that the series begins above the expenditure series and appears to increase massively.

Is there any use of the double y-axis format that is reasonable? My answer is a very restricted "yes." Sometimes the same dependent variable can be represented in a transformed way. For example, plot *log of per pupil expenditures* on the left and *per pupil expenditures* on the right; the latter spaced to match the left-hand scale. Thus those for whom *log dollars* isn't helpful can look on the other scale. Similarly plots of population size against age can be augmented with an axis parallel to the age axis labeled by year of birth. Ironically, no graphics package I know of allows this latter use to be done easily, whereas the misuse is often touted as a desirable option. Alas!

Additional Reading

Report of the Advisory Committee to the Surgeon General of the Public Health Service (1964), *Smoking and Health*. Public Health Service Publication, No. 1103. Washington, D.C., U.S. Government Printing Office.

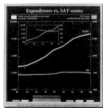

Figure 1.

Figure 2.

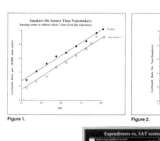

Figure 3. Reprinted by permission of *Forbes* magazine, May 14, 1990. © Forbes Inc., 1990

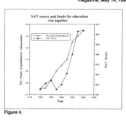

Figure 4.

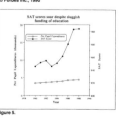

Figure 5.

This may be the most efficient way to pack the whole article onto only two pages, but is getting it printed into the smallest space the point? Fred Mosteller,[1] in a different context, has reminded us that the purpose of instruction is not to cover the topic but to uncover it. It is reminiscent of Einstein's famous remark, *"Everything should be as simple as possible, but no simpler."* Papers should take up as little room as possible, but no less. Surely there must be a better way.

How do we integrate figures and text? The goal should be to have the two intermingled so that they form a single perceptual unit. This is obviously done badly if the figure and its associated text are separated by several pages. It is also done badly if either figure or text have to be so shrunk that their visibility is compromised. Color figures, when allowed at all, often suffer an even more grievous segregation, being isolated in their own signature.

In early texts (e.g., Leonardo's notebooks) no distinction was made between figures and text; the two flowed around one another (see figure 2).

But Gutenberg and the development of movable type destroyed that possibility. Two different processes were involved, and type was set with holes left (usually at the top or bottom of the page) for figures. This limitation has survived into modern times. Most authors know that when preparing a manuscript they must leave an indication for the approximate insertion point for figures, i.e.,

----

Insert a figure about here

----

and the actual "camera ready" figures are isolated at the end of the manuscript. In Jean Paul Richter's otherwise elegant translation of Leonardo's *Notebooks*[1] the figures are reproduced, but the natural flow of text around them is replaced by a squared-off, typeset version of the original Italian and its associated translation.

With the widespread availability of high-resolution laser printers, in which figures and text are treated equally, the continuation of such practice is anachronistic. Greater freedom of placement is readily at hand. How can we use best this newfound freedom? For advice I turned to professionals. The *Information Design Journal* is dedicated to exploring effective ways of conveying information. Not only does this British journal offer advice on how to convey information effectively, it also tries to practice what it preaches. While it offers advice on many fronts (i.e., "left-justify text, but leave right ragged"), I will mention only one here—the use of a wider-than-high format. Libraries hate this because such volumes either stick out gracelessly from their taller companions or hide their bound end and hence their titles.

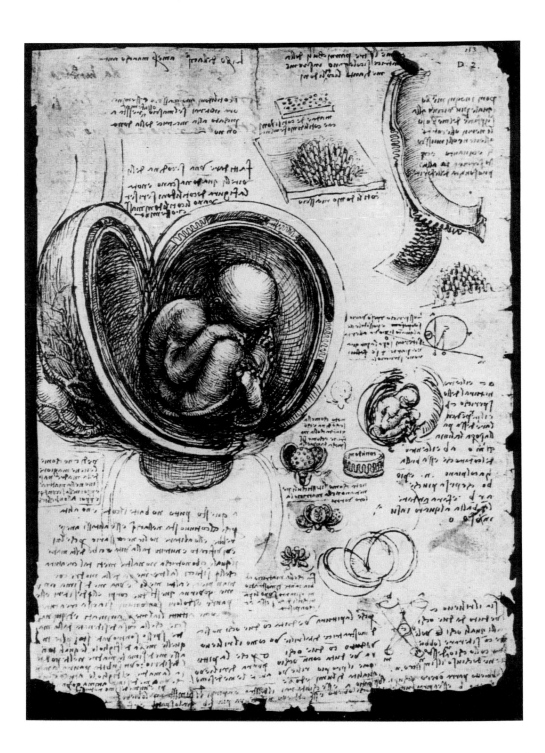

Edward Tufte, who is usually uncompromising in his quest for clarity of presentation, has adopted the traditional higher-than-wide format. Tufte accomplishes the same end through the "inefficiency" of including a lot of white space and so presents a display and its associated description together and then skips to the next page for the text of the next display rather than starting the text prematurely.

FIGURE 2. A page from Leonardo's *Notebooks*, "Studies of Embryos," c. 1510–13, pen over red chalk. (Reproduced with permission from the Windsor Castle, Royal Library. © Her Majesty Queen Elizabeth II.)

CHAPTER 18 Elegance, Grace, Impact, and Graphical Displays

The popular media are filled with poorly designed and poorly executed displays of often important data. A sampling of these are reproduced in section I along with suggestions as to how they might be improved. Many have railed against the abuses of the wonderfully effective tool that graphical display can be and/or have provided guidance for their improved use.[1] Yet bad graphs persist and proliferate. Why?

I had hoped that as microcomputers and associated graphics software became widespread, graphic practice would improve simply because of well-chosen default options. Thus, I reasoned, when the computer-graphics program was set at maximal stupidity (of the user not the machine), a graph of tolerably good quality would emerge. I assumed that there would still be enough built-in flexibility to allow the user to do something terrible, but that one would have to wring the software's neck to accomplish such a nefarious end. Alas, exactly the opposite seems to have occurred. Thus, part of the reason for the continued low quality of media graphics is that the default options of most software packages do not abide by established canons of good practice. But the question remains, why? It must be because software developers think that users like/want these sorts of graphics. Why do they believe this? Probably because users do like them. Why? So far it looks like turtles all the way down.

The ultimate answer lies in pizzazz. I looked at some recent issues of *USA Today, Newsweek, Time,* and some corporate reports. In them I found three-dimensional pies and bars, data that aren't, and colors with no purpose; displays whose primary purpose was surely decoration and not communication. Next I turned to Bill Cleveland's graphical Strunk and White, *The Elements of Graphing Data,* and extracted a good graph (figure 1). It exemplifies the principles of clarity and simplicity that Cleveland emphasizes. A quick look will inform the viewer of the reversal in the trend of bald eagle hatchlings since the banning of DDT. I find this graph both clear and eloquent. Yet it lacks pizzazz.

Must pizzazz be sacrificed if we are to have clarity? I don't think so. William Playfair, the most influential of graphical innovators, desired not only to tell a story graphically but also to tell it dramatically. Tukey[2] echoes this sentiment when he reminds us that *"The greatest possibilities of visual display lie in vividness and inescapability of the intended message."* Impact is important.

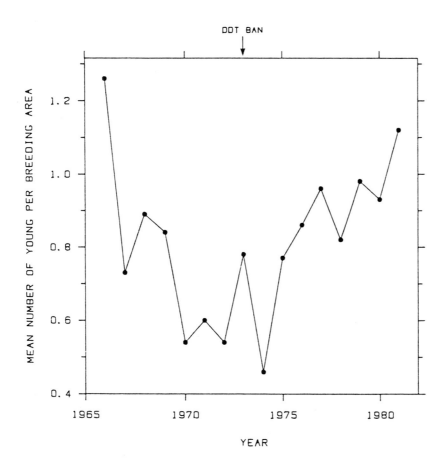

FIGURE 1. A clear graphical depiction of important data from the first edition of Cleveland's *The Elements of Graphing Data,* his figure 2.8.

How can we maintain clarity but add impact? It seems that we must go beyond the Bauhaus principle of minimalism that forms the epistemological basis of Edward Tufte's "data/ink ratio." But even Tufte doesn't believe in blind adherence to such principles. He points out, "The principles [of graphics] should not be applied rigidly or in a peevish spirit; they are not mathematically or logically certain; and it is better to violate any principle than to place graceless or inelegant marks on paper."[3]

Obviously elegance and grace are not new principles, but they are components of what makes a graph memorable and what gives it impact.

How can we give graphs impact without sacrificing clarity? Obviously, the most important component of a memorable graph is the information it contains. I am reminded of what the great blues musician Blind Lemon Jefferson said when asked why there were so few white bluesmen: *"Knowin' all the words in the dictionary ain't gonna help if you got nuthin' to say."* Minard's famous map depicting Napoleon's Russian campaign (chapter 4, figure 1) is evocative partially because the depiction is clear, but more because it dramatically chronicles the death of more than 400,000 soldiers. An identical plot showing the decline in popularity of the Hula-Hoop is unlikely to make anyone's list of the best graphs of all time.

Moving from content to form, what can we do to give our displays more impact without detracting from their clarity? Important instruction in this regard is provided by looking backwards at the great displays in the past. Many of Playfair's graphs, flawed by modern standards, still stand out as works of art. Figure 2 is Playfair's 170-year-old plot[4] showing the relationship between wages and goods over a 260-year period, as well as my attempt to regenerate it (figure 3). My version was produced in under an hour with readily available Macintosh software.

FIGURE 2. William Playfair's 1821 *Letter on Our Agricultural Distresses,* which shows that over the previous 260 years (Playfair apparently believed in the importance of examining long-term trends) the cost of wheat rose faster than the wages of a good mechanic.

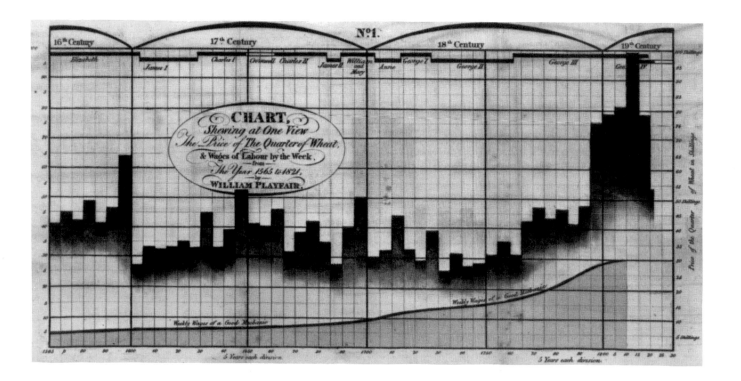

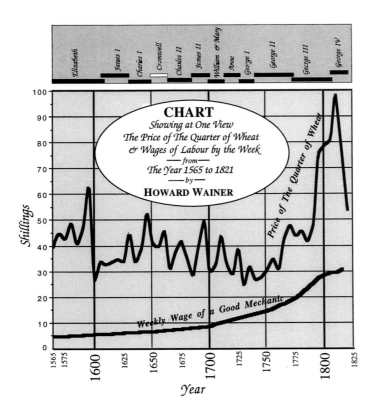

FIGURE 3. A version of Playfair's chart produced with modern computer graphic software in under an hour by a user with only moderate skill.

I find this result both visually pleasing and not too overburdened with chartjunk to confuse the message contained in the data. Thus encouraged, I tackled Cleveland's bald eagles (figure 4). I leave it to the eye of the reader to judge the extent to which this represents a real improvement.

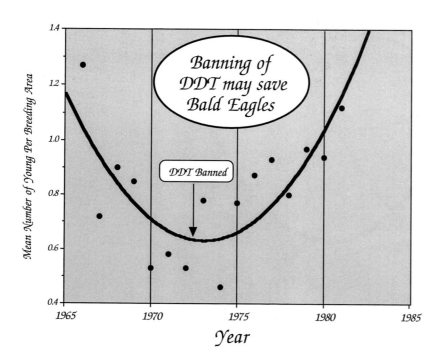

FIGURE 4. A modest attempt at using what we learn from Playfair to redraw Cleveland's bald eagle graph with a little more pizzazz while maintaining the clarity of its message.

Does this same approach work with more technically sophisticated plots? If Playfair were alive today, were a baseball fan, and knew about Quantile plots, how would he compare the batting averages of National with American League hitters? Perhaps something like figure 5 is what might have resulted.

FIGURE 5. An attempt to meld a complex statistical idea with an older format. Each point on the plot represents a specific point in both leagues (say the batting average at which twenty percent of the players were better). The many points represent many different percentages. A point plotted on the diagonal line means that a player with that average has the same rank among the other batters of his league as he would in the other league. The fact that most points are above the diagonal means that National League hitters at all levels did better than those in the American League.

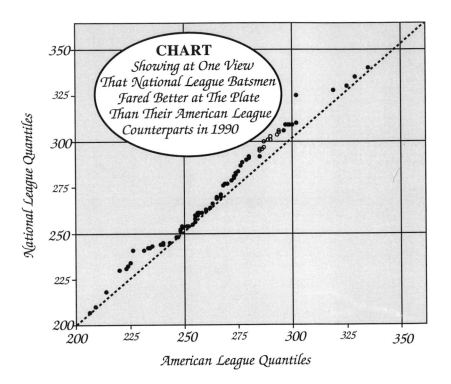

Obviously, I have used the terms "elegance" and "grace" in a somewhat broader sense than the traditional mathematical meaning, which usually connotes a kind of minimal leanness. This same sort of leanness is also used when describing the elegance of such minimal structures as suspension bridges.

Yet we have learned that even rococo Victorian houses carry a grace and elegance on their own terms. Austerity may serve certain purposes, but humans often prefer, even require, more. We need to pay attention to experience and to the results of careful experiments, but we also need to heed the lore of graphic designers. Perhaps there is something yet to be learned from the success enjoyed by the multi-colored, three-dimensional pie charts that clutter the pages of *USA Today, Time,* and *Newsweek.*

I sure hope not much.

CHAPTER 19 Sense-Lining

She is older than the rocks among which she sits; like the vampire, she has been dead many times, and learned the secrets of the grave; and has been a diver in deeps seas, and keeps their fallen day about her; and trafficked for strange webs with Eastern merchants; and, as Leda, was the mother of Helen of Troy, and as Saint Anne, the mother of Mary; and all this has been to her but as the sound of lyres and flutes, and lives only in the delicacy with which it has moulded the changing lineaments, and tinged the eyelids and the hands.

From Walter Pater, *Studies in the Renaissance*

She is older than the rocks among which she sits;
Like the vampire,
She has been dead many times,
And learned the secrets of the grave;
And has been a diver in deeps seas,
And keeps their fallen day about her;
And trafficked for strange webs with Eastern merchants;
And, as Leda,
Was the mother of Helen of Troy,
And as Saint Anne, the mother of Mary;
And all this has been to her but as the sound of lyres and flutes,
And lives
Only in the delicacy
With which it has moulded the changing lineaments,
And tinged the eyelids and the hands.

From Walter Pater, "Mona Lisa,"
in *The Oxford Book of Modern Verse 1892–1925.*

This example appeared in a paper by George Cuming[4] that I recently came across advocating "sense-lining" in liturgical typography. Cuming's point was that since difficulties of reading aloud are eased by breaking prose at points that make sense rather than at the end of a line, it was not implausible that difficulties in understanding could be eased using the same strategy. It seems to me that precisely the same argument could be applied to the way that we print in many other circumstances.

For example, the IRS provides the following instructions on page 17 of the all too familiar booklet describing the 1991 form 1040:

Lines 13 and 14

Capital Gain or (Loss)

Enter on line 13 your capital gain or (loss) from Schedule D. If you received **capital gain distributions** (reported to you on **Form 1099-DIV** or a substitute statement) but do not need Schedule D for other capital transactions, enter those distributions on line 14.

Caution: *It will be to your advantage to report your capital gain distributions on Schedule D and use Part IV of Schedule D to figure your tax if your taxable income (Form 1040, line 37) is **more than:** $82,150 if married filing jointly or qualifying widow(er); $49,300 if single; $70,450 if head of household; or $41,075 if married filing separately.*

A modest attempt at revision that includes sense-lining (and a little reorganization) might yield:

Lines 13 and 14

Capital Gain or (Loss)

If you received **capital gain distributions** and if you
- **Use Schedule D,** enter your capital gain or (loss) on line 13.
- **Do not use Schedule D,** enter those distributions on line 14.

Caution:
If your taxable income (Form 1040, line 37) **is more than:**
   $82,150 if married filing jointly or qualifying widow(er);
   $70,450 if head of household;
   $49,300 if single; or
   $41,075 if married filing separately,
it will be to your advantage to report your capital gain distributions on Schedule D and use Part IV of Schedule D to figure your tax.

A second example (closer to the liturgical beginnings of this notion) is from the instructions that are given to examinees for a standardized test. These are usually read aloud by the test administrators at the same time that the examinees read them silently. Shown below are the instructions from the advanced placement chemistry exam. These are typical of those for most standardized tests.

> Use your time effectively, working as rapidly as you can without losing accuracy. Do not spend too much time on questions that are too difficult. Go on to other questions and come back to the difficult ones later if you have time. It is not expected that everyone will be able to answer all the multiple-choice questions.

Once again, sense-lining (a first attempt below) can convey better the meaning of the prose to the reader.

U se your time effectively,
working as rapidly as you can without losing accuracy.

Do not spend too much time on questions that are too difficult.

Go on to other questions
       and
come back to the difficult ones later if you have time.

Many will not have time to answer all the multiple-choice questions.

Perhaps with experienced test-takers such niceties aren't important, but I suspect that where experience is less common, clearer instructions would be helpful. Moreover, in assessments in lower grade levels or assessment tasks that are less familiar in form than the traditional multiple-choice test, the efficacy of typographic simplifications such as these ought to be explored.

Sense-lining is not a totally new idea. Paragraphs have much the same object, but sense-lining carries the principle much further. Most would agree that the appreciation and comprehension of verbal material is crucially dependent upon its proper articulation. The extra step of sense-lining is to presume the importance of this articulation in its printed representation.

Does sense-lining take up too much space? I don't believe this must be so. It may not even take up any more space. Cuming[2] extracted an illustration of this from *The Alternative Service Book* "of The Episcopal Church in America." Showing that the two-line original,

> The law was given by Moses: grace and truth came through
> Jesus Christ. *John 1.17*

is no more parsimonious than a sense-lined alternative,

> The law was given by Moses:
> grace and truth came through Jesus Christ. *John 1.17*

But even in those situations in which sense-lining and other typographical techniques are not as spatially parsimonious as "filling the page from margin to margin,"

who is to say (without experimental data) that even if sense-lining takes up more space it is still not, em-for-em, more efficient at communicating than would have been the case in the more traditional mode.

CHAPTER 20 Making Readable Overhead Displays

I have just awakened from an uncomfortable nap in a lecture room down the hall from my office. My loss of consciousness was unintended. I had expected to learn about some new methods for detecting poorly performing test items. The speaker tried to use overhead transparencies to illuminate his ideas. Unfortunately, the transparencies helped to obscure the vast darkness of his topic. Their unintelligibility, combined with the darkened lecture hall, contributed mightily to my sleepy condition. The fact that his transparencies were largely unreadable was less disturbing to me than how similar his transparencies were to displays that I see in most talks. Few complaints are heard because we are inured to such woeful inadequacy by the dreadful norms. Although I have often railed against such practice, all that it has gotten me is a perfunctory remark that the speaker will often make upon spotting me in the audience on the order of "Howard will hate my overheads"; then they go on anyway.

Why do people of reasonable intellectual capacity continue to produce overhead displays that are entirely unreadable and hence worthless for their purpose? Figure 1 is an all too familiar example, obviously generated by photocopying some unedited computer output onto a transparency.

The display is overcrowded; the printing is too small; the text is full of jargon; the numbers are overburdened with insignificant digits and irrelevancies. It is worse than unintelligible; it is ugly and graceless.

How can we improve matters? As always, it is useful to start with an explicit statement of purpose. The principal goal of any presentation is communication. To effectively communicate we need to make some assumptions about the audience, but in addition, we need to understand the medium within which the communication is to take place. I have found it sensible to take a conservative (pessimistic) attitude toward both of these aspects. In particular, I assume that I am trying to provide a clear view of my material to someone of poor eyesight, sit-

```
***** SIMULTANEOUS ITEM BIAS ESTIMATION and HYPOTHESIS TESTING *****    a1
Theory by R. Shealy and W. Stout, program by L. Roussos and R. Shealy

number of items on test = 36
name of file for Ref. grp. scores = histmal.cvl
name of file for Focal grp. scores = histfem.cvl
minimum no. of examinees per statistic calculation cell =    2
estimate of guessing on the test = 0.20
number of runs for this data set = 36
number of examinees in Reference group = 499
number of examinees in Focal group       = 500


         p-value notation:
R    denotes p-value for test of bias/DIF/DTF against Ref. group
F    denotes p-value for test of bias/DIF/DTF against Foc. group
E    denotes p-value for test of bias/DIF/DTF against either the
                                              Ref. or Foc. groups.

NOTE:   M-H Chi-square p-value is restricted, by definition, to type E.
```

				Mantel-Haenszel		
run no.	Beta-uni	SIB-uni z-statistic	SIB-uni p-value	Chi sqr.	p value	Delta (D-DIF)
1	0.109	3.789	0.000 E	12.81	0.000 E	-1.34
2	-0.043	-1.441	0.150 E	2.17	0.140 E	0.54
3	0.037	1.445	0.148 E	1.71	0.192 E	-0.55
4	-0.012	-0.396	0.692 E	0.14	0.709 E	0.15
5	-0.033	-1.153	0.249 E	1.05	0.306 E	0.39
6	0.027	0.896	0.370 E	0.53	0.467 E	-0.27
7	0.002	0.078	0.938 E	0.01	0.939 E	-0.06
8	-0.021	-1.439	0.150 E	0.75	0.386 E	0.65
9	-0.078	-2.903	0.004 E	6.46	0.011 E	1.00
10	-0.002	-0.075	0.940 E	0.06	0.807 E	0.11
11	-0.085	-2.821	0.005 E	7.37	0.007 E	0.95
12	0.112	4.332	0.000 E	16.91	0.000 E	-1.69
13	-0.018	-0.901	0.368 E	1.14	0.286 E	0.59
14	0.041	1.355	0.175 E	1.40	0.237 E	-0.42
15	0.097	3.705	0.000 E	11.44	0.001 E	-1.33
16	0.012	0.752	0.452 E	0.00	0.944 E	-0.13
17	0.030	1.102	0.270 E	1.20	0.274 E	-0.46
18	0.029	1.008	0.313 E	1.40	0.237 E	-0.45
19	0.016	0.579	0.563 E	0.23	0.630 E	-0.21
20	-0.030	-1.307	0.191 E	0.98	0.321 E	0.47
21	-0.019	-0.632	0.527 E	0.30	0.584 E	0.22
22	-0.132	-4.307	0.000 E	18.57	0.000 E	1.48
23	0.021	0.653	0.514 E	0.27	0.606 E	-0.19
24	-0.078	-2.499	0.012 E	6.19	0.013 E	0.86
25	0.028	0.917	0.359 E	1.36	0.244 E	-0.42
26	-0.019	-0.692	0.489 E	0.25	0.620 E	0.21
27	0.027	1.051	0.293 E	1.46	0.228 E	-0.49
28	-0.006	-0.194	0.846 E	0.44	0.507 E	0.23
29	-0.038	-1.673	0.094 E	3.15	0.076 E	0.88
30	0.095	3.684	0.000 E	11.98	0.001 E	-1.43
31	-0.009	-0.333	0.739 E	0.18	0.668 E	0.19
32	0.039	1.391	0.164 E	2.77	0.096 E	-0.65
33	0.015	0.486	0.627 E	0.06	0.812 E	-0.10
34	-0.098	-3.482	0.000 E	11.53	0.001 E	1.28
35	-0.014	-0.507	0.612 E	0.00	0.950 E	0.05
36	0.022	0.781	0.435 E	0.53	0.466 E	-0.31

Exploratory Stage

```
Program execution is completed.

Your output is stored on the file:  explore.out
```

FIGURE 1. An example of an all too typical unreadable overhead display

ting at the back of a long and smoke-filled hall, on a too-small screen, using an aging projector of inferior quality.

With these restrictions in mind, I have found the following guidelines useful.

1. **Center all material in a 6" by 9" frame** in the middle of the 8.5" by 11" transparency. Doing so reduces the impact of the spherical distortion that appears around the edges of inferior projectors. It also helps reduce the temptation to squeeze on more stuff.

2. **Use no more than thirty characters per line.** This is a pretty severe restriction. Note that the advice itself used forty characters.

3. **Use no more than fifteen lines per overhead.** This is really an upper bound. Fewer lines may be called for when the material is dense; fifteen lines of mathematics should almost surely be broken up (probably into three talks).

4. **Use 36-point type for major headings, 24-point for the rest.** These are somewhat redundant with guidelines 1–3. If you adhere to guidelines 1 and 2 while making the type size as large as possible within those guidelines, you will discover that 24-point type will only allow about thirty-six characters in a line and 36-point only twenty-four characters.

Figure 2 shows an overhead display that satisfies these rules. It is not as full as it could be, and more lines could have been added, but it does illustrate the amount of self-control that is required for producing good overheads.

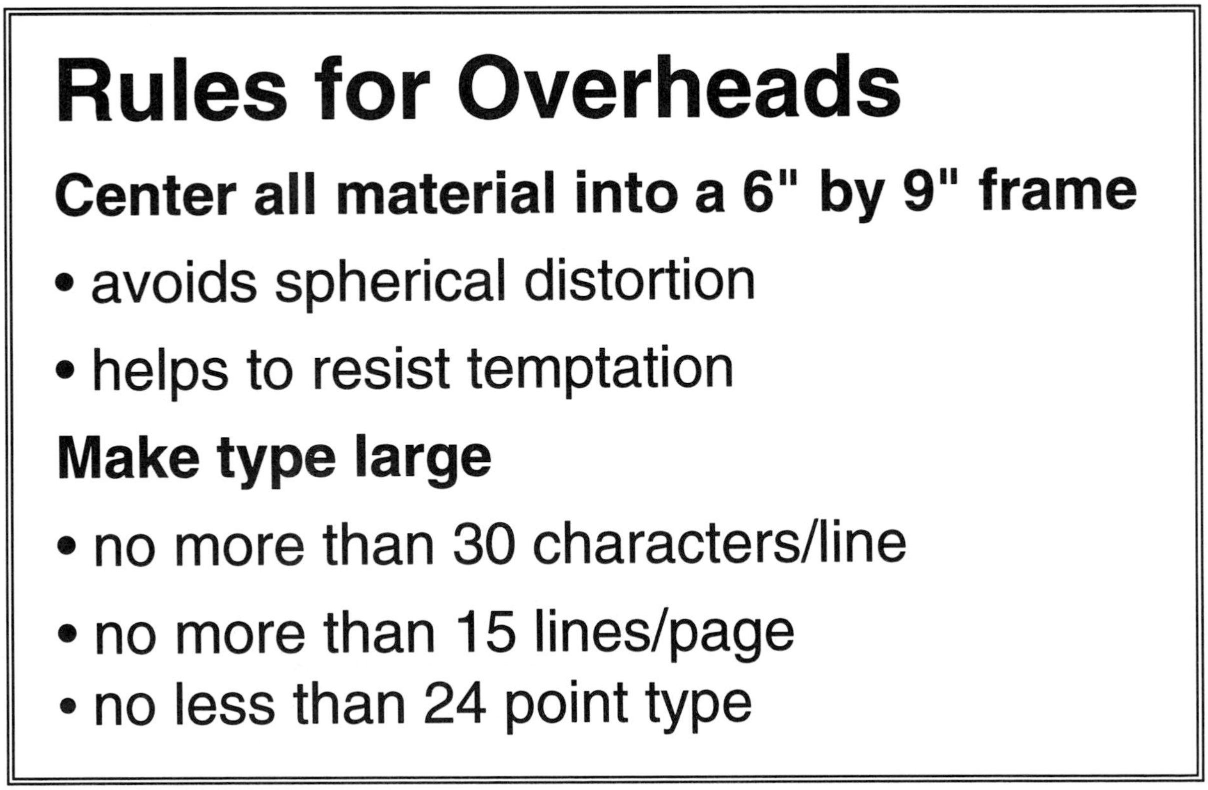

Rules for Overheads

Center all material into a 6" by 9" frame

- avoids spherical distortion

- helps to resist temptation

Make type large

- no more than 30 characters/line

- no more than 15 lines/page

- no less than 24 point type

Is this all that needs to be done to assure good overheads? No. But it will assure readable overheads. There are a few other ideas that will also increase the readability of overheads, but they are likely to have a more modest effect.

FIGURE 2. The basic rules for making overheads in a format that satisfies rules.

5. **Limit the number of fonts used.** Two is usually plenty. Modern word processors provide enormous flexibility in this regard.

Discipline is critical to avoid "fontification." Pick a clear sans serif font (Helvetica is a safe choice) and stay with it. Use boldface discreetly. Don't use "all caps" for text.

6. **Resist seduction by color.** Color is pretty but only rarely helpful for much more than emphasis ("Note the red points"). Color rarely helps communication, and its unwise use can harm legibility; some colors simply disappear under less than optimal viewing conditions. Stick with black and white as much as possible.

7. **Remember that these rules have limited flexibility.** Is your overhead likely to be useless if you have thirty-three characters on a line? If you use 18-point type? If you have sixteen lines? No. But each step past these boundaries limits the robustness of your overheads. If the room you are speaking in is better than the one proposed here initially, you can get away with it. You will be helped if the audience is sharp-eyed and sits close to the front. To know how flexible these rules are, it is important whenever possible to go to the room in which you will be giving the talk and try your overheads out. Put your worst display on the projector and then see how far back you can sit and still read it without squinting. If you can't read it when you are against the back wall, either redo the ineffectual overheads or have as many of the back rows of chairs removed as necessary.

8. **Limit the number of equations you present** to a maximum of two or three per overhead, fewer if they are complex. It is probably wise (depending upon the audience) to limit the total number of equations to five per hour. Few audiences can follow more than that.*

9. **Limit the number of digits shown** in any numbers to two significant figures. Audiences rarely need, want, or understand more. See chapter 10 for an expansion and explanation of this advice.

I hope that this codification is helpful. Underlying all of these rules is some common sense. When preparing for a talk, construct the overheads carefully. Just because you can photocopy your typed paper onto a transparency does not mean that it is a good idea to do so. It isn't.

Next, if you can't try out your transparencies under the actual conditions (guideline 7), do so under parallel (or worse) conditions. If you can't read them from the back of the room, redo the transparencies. Rehearse the talk and notice what you say. If the number on the screen is $27,341,856.22 but you say "we spent about $27 million dollars," round the figure on the transparency to two digits.

When using transparencies, platform performance is improved if when addressing the material on the transparency, you point to the screen and not to the transparency. This has three immediate advantages:

*One talk by an eminent economist given at the Educational Testing Service some years ago was so full of equations that John Tukey said they "made his eyes bug out" (see Wainer, 1986, p. 58). Tukey can deal with mathematical equations "on the fly" about as well as anyone ever has. If he couldn't follow, who was the speaker addressing?

- It forces you to look at the screen and hence reduces the likelihood that the material will be displayed out of focus, half off the screen, or upside down.
- It provides a less obstructed view from the audience. More people can see what you are pointing at. Moreover, it prevents you from displaying half the material of the transparency on the shoulder attached to the finger you are pointing with.
- It eliminates the irritating ersatz palsy that is generated when the projector magnifies the slightest jiggle of your nervous hand.

Last, how can the perpetrator of the transparency in figure 1 effectively display his 36-by-8 matrix of numbers? He can't. What's more, a realization of the impossibility of such a depiction should lead to a rethinking of what really needs to be displayed. Perhaps a verbal summary. Perhaps a graph. Remember that in the Gettysburg Address Lincoln needed to report only a single number, and nothing we might say is likely to be as important.

Finally—

In this book I have reproduced some of the finest examples of effectively displayed information; also some of the worst. The best of these displays always seem to have been constructed in a minimalist style reminiscent of Japanese interior decoration. The message is featured and not the container. Minard's metaphorical depiction of the river of Napoleon's army flowing across the Urals until being dammed by the Russian winter (chapter 4, figure 1) is perhaps the best example. There are many others. No one who has seen and digested the meaning of the two lines of Norman Maclean's graphical recounting of the race between young men and fire (chapter 4, figure 3) can forget the tragic consequences of their intersection. The catastrophic implications of launching the *Challenger* if the temperature drops to near freezing shout at us from the elegant spareness of a single swooping line in figure 7 of chapter 2.

The human ability to decode graphics is so great that important messages can sometimes still be discerned within a chart overloaded with visual irrelevancies, but the clarity of the message depends, critically, on the simplicity of the display. A byzantine design is usually a substitute for important information, not an accompaniment. Designers often make the form of a chart complex to hide its inadequacies of content. Sadly, no amount of graphical ability will allow the extraction of information from a display when that information was not included in its construction. Images of silk purses and pig ears are appropriate.

In 1917, T.E. Lawrence, only just beginning his adventures in Arabia, became the British Army's liaison to Prince Feisal. He observed that the phrases that the prince chose to express himself

> were usually the simplest.... It seemed possible, so thin was the screen of words, to see the pure and very brave spirit shining out."[1]

This is a remarkably apt characterization of effective visual communication.

Notes

Introduction
1. Fletcher (1849).
2. Three important books in this category are Bertin (1973), Cleveland (1994), and Schmid and Schmid (1979).
3. Tufte (1983, 1990, 1997).

Section I
1. *Smoking and Health: Report of the Advisory Committee to the Surgeon General of the Public Health Service.* Public Health Service Publication No. 1103. U.S. Government Printing Office: Washington, DC, 1964.

Chapter 1
1. Tufte (1983).
2. Friedman and Rafsky (1981).
3. Kelley, Ayres, and Bowen (1967).
4. Crotty (1970).
5. Tufte (1970).
6. This example is drawn from Cleveland and McGill (1984).
7. Wainer and Francolini (1980).
8. von Mayr (1874).
9. Bertin (1973).
10. Tufte (1983, 1990, 1997).

Chapter 2
1. Thompson (1782).
2. Konner and Worthman (1980)
3. Presidential Commission on the Space Shuttle *Challenger* Accident (1986), p. 145.

Chapter 3

1. Tilling (1975), p. 193.
2. Laitman and Reidenberg (1992).
3. Culotta (1993)
4. Mangel and Samaniego (1984); Wald (1980).

Chapter 4

1. Marey (1885).
2. Tufte (1980), p. 40.
3. Maclean (1992).
4. Ibid.

Chapter 5

1. Biderman (1978).
2. Feynman (1985), p. 85.
3. Durso and Wainer (1996).
4. Feynman (1949b), p. 769.

Chapter 7

1. Guthrie, Seifert, and Kirsch (1986).

Chapter 8

1. American Psychological Association (1994), p. 143.

Chapter 10

1. Farquhar and Farquhar (1891).
2. This type of graphic is developed and described in Tukey (1977).
3. Wainer and Schacht (1978).

Chapter 11

1. Fienberg (1975).
2. Ashford and Sowden (1970).
3. Funkhouser (1937); Lalanne (1845).
4. See de Barros de Santarem (1895) for sketches of Bianco's map of 1448.

Chapter 12

1. Playfair (1786).
2. Guilbaud (1946).
3. Coleman (1961), p. 14.
4. Schmid and Schmid (1979), p. 151.

5. Upton (1994).
6. Ibid.

Chapter 13

1. Bertin (1973), p. 109.

Chapter 14

1. Policy Information Center (1990).

Chapter 16

1. Biderman (1963).
2. Salganik, Phelps, Bianchi, Nohara, and Smith (1993).
3. Mullis, Dossey, Owen, and Phillips (1993).
4. Kepler (1609).

Chapter 17

1. Richter (1883).

Chapter 18

1. For example, Bertin (1973), Cleveland (1985), Tufte (1983),
   Tukey (1977, 1990), Wainer (1980, 1984).
2. Tukey (1990), p. 328.
3. Tufte (1983), p. 191.

Chapter 19

1. Cuming (1990).
2. Ibid, p. 92.

Finally—

1. Lawrence (1935), p. 123.

References

American Psychological Association. (1994). *Publication Manual of the American Psychological Association* (4th ed.). Washington, DC.

Arbuthnot, J. (1710). "An argument for Divine Providence taken from the Constant Regularity in the Births of Both Sexes." *Philosophical Transactions of the Royal Society, London* 27, 186–190.

Arensburg, B., Tillier, A.M., Vandermeersch, B., Duday, H., Schepartz, L.A., and Rak, Y. (1989). "A middle Palaeolithic human hyoid bond." *Nature* 338, 758–760.

Ashford, J.R., and Sowden, R.D. (1970). "Multivariate probit analysis." *Biometrics* 26, 535–546.

Baude, M. (1847). "Rapport fait à la commision spéciale chargée de rechercher les moyens de sûreté applicables aux chemins de fer sur le régulateur des chemins de fer de M. IBRY." *Annales des Ponts et Chaussées*, XIII, 2e Série, 200–204.

Bertin, J. (1973). *Semiologie Graphique* (2d ed.). The Hague: Mouton-Gautier. (English translation by William Berg and Howard Wainer, published as *Semiology of Graphics*, Madison, WI: University of Wisconsin Press, 1983.)

Bertin, J. (1977). *La Graphique et le Traitement Graphique de L'Information.* Paris: Flammarion.

Bertin, J. (1980). *Graphics and the Graphical Analysis of Data.* (Translation prepared by W. Berg and H. Wainer). Berlin: De Gruyter.

Biderman, A.D. (1978). *Intellectual impediments to the development and diffusion of statistical graphics, 1637–1980.* Presented at 1st General Conference Social Graphics, Leesburg, VA.

Biderman, A.D., Louria, M., and Bacchus, J. (1963). "Historical incidents of extreme overcrowding." Technical Report 354–5, Washington, DC: Bureau of Social Science Research.

Brakenridge, W. (1755). "A letter from the Reverend William Brakenridge, D.D. and F.R.S. to George Lewis Scot, Esq.,

F.R.S., concerning the *London Bills of Mortality.*" *Philosophical Transactions of the Royal Society, London* 48, 788–800.

Chamberlain, T.C. (1965). "The method of multiple working hypotheses." *Science* 148, 754–759.

Cheng, P.C-H., and Simon, H.A. (1995). Scientific discovery and creative reasoning with diagrams." In S. Smith, T. Ward, and R. Finke (Eds.), *The Creative Cognition Approach.* Cambridge, Mass.: MIT Press.

Cleveland, W.S. (1994). *The Elements of Graphing Data.* (2d ed.): Summit, NJ: Hobart Press.

Cleveland, W.S., and McGill, R. (1984). "Graphical perception: theory, experimentation, and application to the development of graphical methods." *Journal of the American Statistical Association* 79, 531–554.

Coleman, J. (1961). *Social Climates in High Schools.* Washington, D.C.: Government Printing Office.

Costigan-Eaves, P. (1984). *Data Graphics in the 20th Century: A Comparative and Analytic Survey.* (unpublished doctoral dissertation), New Brunswick, NJ: Rutgers University.

Cox, D.R. (1978). "Some remarks on the role in statistics of graphical methods." *Applied Statistics* 29, 4–9.

Crotty, W.J. (ed.) (1970). *Public Opinion and Politics: A Reader.* New York: Holt, Rinehart and Winston, p. 364.

Culotta, E. (1993). "At each others' throats." *Science* 260, 893.

Cuming, G. (1990). Liturgical typography: a plea for sense-lining. *Information Design Journal* 6, 89–92.

Dalal, S.R., Fowlkes, E.B., and Hoadley, B. (1989). "Risk analysis of the space shuttle: Pre-*Challenger* prediction of failure." *Journal of the American Statistical Association* 84, 945–957.

de Barros de Santarem, M.F. (1895). "Pre-Columbian Discovery of America." *Geographical Journal* 5,(3), 222, 224, 226.

Durso, J., and Wainer, H. (1996). "A Nobel Plot." *Chance* 9(1), 12–16.

Eddy, W.F. (1990). "Review of *Envisioning Information* by E.R. Tufte." *Chance* 3(3), 60–61.

Ehrenberg, A.S.C. (1977). "Rudiments of numeracy." *Journal of the Royal Statistical Society,* A, 140, 277–297.

Farquhar, A.B., and Farquhar, H. (1891). *Economic and Industrial Delusions: A Discourse of the Case for Protection.* New York: Putnam.

Feynman, R.P. (1949a). "The theory of positrons." *The Physical Review* 76, 749–759.

Feynman, R.P. (1949b). "Space-time approach to quantum electrodynamics." *The Physical Review* 76, 769–789.

Feynman, R.P. (1985). *QED.* Princeton: Princeton University Press.

Fienberg, S.E. (1975). "Perspective Canada as a social report." *Social Indicators Research* 2, 153–174.

Fletcher, J. (1849). "Moral and educational statistics of England and Wales." *Journal of the Statistical Society of London* 12, 151–176, 189–335.

Friedman, J.H., and Rafsky, L.C. (1981). "Graphics for the Multivariate Two-Sample Problem." *Journal of the American Statistical Association* 76, 277–287 .

Funkhouser, H.G. (1937). "Historical development of the graphic representation of statistical data." *Osiris* 3, 269–404.

Gilbert, E.W. (1958). "Pioneer maps of health and disease in England." *Geographical Journal* 124, 172–183.

Graunt, J. (1662). *Natural and Political Observations on the Bills of Mortality.* London: Martyn.

Guilbaud, G.Th. (1946). "Méthode d'analyse sommaire de la structure démographique." *Economie et humanisme* 22, 515–525.

Guthrie, J.T., Seifert, M., and Kirsch, I.S. (1986). "Effects of education, occupation, and setting on reading practices." *American Educational Research Journal* 23, 151–160.

Jaret, P. (1991). "The disease detectives." *National Geographic* 179, 114–140.

Joint Committee on Standards for Graphic Presentation. (1915). "Preliminary Report." *Journal of the American Statistical Association* 14, 790–797.

Kelley, S., Ayres, R.E., and Bowen, W.G. (1967). "Registration and voting: putting first things first." *American Political Science Review* 61, 371–385.

Kepler, J. (1609). Astronomia nova. In M. Caspar (ed.) *Gesammelte Werke,* 3, München: Beck. 1937.

Kirsch, I.S., and Jungeblut, A. (1988). *Literacy: Profiles of America's young adults.* NAEP Report 16-PL-02. Princeton, NJ: Educational Testing Service.

Kirsch, I.S., Jungeblut, A., Jenkins, L., and Kolstad, A. (1993). *Adult Literacy in America.* Washington, DC: National Center for Education Statistics.

Konner, M., and Worthman, C. (1980). "Nursing frequency, gonadal function and birth spacing among !Kung hunter-gatherers." *Science* 207, 788–791.

Laitman, J., and Reidenberg J. (1992). "Neanderthal language debate: tongues waganew." *Science* 3 April, p. 33.

Lalanne, L. (1845). "Sur la représentation graphique des tableaux météorologiques et des lois naturelles en général." Appendix to *Cours complet de météorologie de L. F. Kaemtz, traduit et annoté par Ch. Martins,* Paris.

Lawrence, T.E. (1935). *Seven Pillars of Wisdom.* Garden City, NY: Doubleday, Doran & Co.

Maclean, N. (1992). *Young Men and Fire.* Chicago: University of Chicago Press.

Mangel, M., and Samaniego, F. J. (1984). "Abraham Wald's work on aircraft survivability." *Journal of the American Statistical Association* 79, 259–267.

Marey, E.J. (1885). *Le Méthode Graphique.* Paris: Boulevard Saint Germain et rue de l'Eperon.

Margerison, T. (1965). "Review of Writing Technical Reports by Bruce M. Copper." *London Times,* 3 January. Quoted on p. 49 of *A Random Walk in Science,* R.L. Weber, compiler, E. Mendoza (ed.). London: Institute of Physics; New York: Crane, Russak.

Maury, M.F. (1848–80). *Wind & Current Chart Series A-B, C-F.* Washington, DC: United States Hydrographical Office.

McIntyre, J., Myint, T., and Curtis, L. (1979). "The effects of situation and mode of resistance on the likelihood of injury and rape during sexual assault." Annual Meeting of the American Sociological Association, Boston, Massachusetts.

Merton, R.K. (1973). *The Sociology of Science: Theoretical and Empirical Investigations.* Edited and with an introduction by Norman W. Storer. Chicago: University of Chicago Press.

Minard, C.J. "Tableaus Graphiques et Cartes Figuratives de M. Minard, 1845–1869."

Mosteller, F. (1980). Classroom and platform performance. *The American Statistician* 34, 11–16.

Mullis, I.V.S., Dossey, J.A., Owen, E.H., and Phillips, G.W. (1993), "NAEP 1992: Mathematics Report Card for the Nation and the States," Report 23-ST02. Washington, DC: National Center for Education Statistics.

Nightingale, F. (1858). *Notes on Matters Affecting the Health, Efficiency and Hospital Administration of the British Army.* London.

Perot, R. (1992). *United We Stand: How We Can Take Back Our Country.* New York: Hyperion.

Playfair, W. (1786). *The Commercial and Political Atlas.* London: Corry.

Playfair, W. (1805). *An Inquiry into the Permanent Causes of the Decline and Fall of Powerful and Wealthy Nations,* London: Greenland & Norris.

Playfair, W. (1821). *A Letter on Our Agricultural Distresses. Their Causes and Remedies,* 1st ed. London: William Sams.

Policy Information Center. (1990). *From School to Work.* Policy Information Report of the Policy Information Center. Princeton, NJ: Educational Testing Service.

Presidential Commission on the Space Shuttle *Challenger* Accident (1986). *Report of the Presidential Commission on the Space Shuttle Challenger Accident* (Volumes 1 & 2), Washington, DC: Author.

Priestley, J (1769). *A New Chart of History.* London.

Richter, J.P. (translator and editor) (1883). *The Literary Works of Leonardo da Vinci.* London: Sampson, Low, Marston, Searle & Rivington.

Rothermel, R. (1972), *Mathematical Model for Predicting Spread and Intensity of Fire in Wildland Fuels.* U.S. Forest Service Report.

Salganik, L.H., Phelps, R.P., Bianchi, L., Nohara, D., and Smith, T.M. (1993). *Education in States and Nations: Indicators Comparing U.S. States with the OECD Countries in 1988.* NCES Report No. 93–237. Washington, DC: National Center for Education Statistics.

Schmid, C.F. (1954). *Handbook of Graphic Presentation.* New York: Ronald.

Schmid, C.F., and Schmid, S.E. (1979). *Handbook of Graphic Presentation* (2nd ed.). New York: Wiley.

Stigler, S.M. (1980). "Stigler's Law of Eponymy." *Transactions of the New York Academy of Sciences, 2nd Series* 39, 147–157.

Thompson, B. (1782). "New experiments upon gun-powder with occasional observations and practical inferences; to which are added an account of a new method for determining the velocities of all kinds of military projectiles and the description of a very accurate Epouvette for gun-powder." *Philosophical Transactions* 71, 229–328.

Thompson, D.W. (1961), *On Growth and Form.* Cambridge, UK: Cambridge University Press.

Tilling, L. (1975). Early experimental graphs. *British Journal of Historical Science* 8, 193–213.

Tufte, E.R. (1970). *The Quantitative Analysis of Social Problems.* Reading, MA: Addison-Wesley.

Tufte, E.R. (1977). Improving data display. Department of Statistics, University of Chicago.

Tufte, E.R. (1983). *The Visual Display of Quantitative Information.* Cheshire, CT: Graphics Press.

Tufte, E.R. (1990). *Envisioning Information.* Cheshire, CT: Graphics Press.

Tufte, E.R. (1997). *Graphical Explanations.* Cheshire, CT: Graphics Press.

Tukey, J.W. (1977). *Exploratory Data Analysis.* Reading, MA: Addison-Wesley.

Tukey, J.W. (1990). "Data based graphics: visual display in the decades to come." *Statistical Science* 5, 327–329.

Upton, G.J.G. (1994). "Picturing the 1992 British General Election." *Journal of the Royal Statistical Society (A)* 157, 231–252.

U.S. Bureau of the Census (1980). *Social Indicators III.* Washington, DC

U.S. Public Health Service (1964). *Smoking and Health:* Report of the Advisory Committee to the Surgeon General of the Public Health Service. Public Health Service Publication No. 1103. Washington, DC: U.S. Government Printing Office.

von Mayr, G. (1874). *Gutachten (Über die Anwendung der graphischen and geographischen Methoden in der Statistik.* Munich.

Wahlen, M., Kunz, C.O., Matuszek, J.M., Mahoney, W.E., and Thompson, R.C. (1980). "Radioactive plume from the Three Mile Island accident: xenon-133 in air at a distance of 375 kilometers." *Science* 207, 639–640.

Wainer, H. (1980). "A timely error." *Royal Statistical Society News and Notes* 7, 6.

Wainer, H. (1980a). "Making newspaper graphs fit to print." In P. Kolers, M.E. Wrolstad, and H. Bouma (eds.). *Processing of Visible Language: 2.* (pp. 125–142). New York: Plenum. (Republished in two parts: Chapel Hill, NC, *Newspaper Design Notebook*, 1981, 2, (6), 1, 10–16, and 1981, 3, (1), 3–5.).

Wainer, H. (1983). "How are we doing? A review of Social Indicators III." *Journal of the American Statistical Association* 78, 492–496.

Wainer, H. (1984). "How to display data badly." *The American Statistician* 38, 137–147.

Wainer, H. (1986). *Drawing Inferences From Self-Selected Samples,* New York: Springer-Verlag.

Wainer, H., and Francolini, C. (1980). "An empirical inquiry into human understanding of 'two variable color maps.'" *The American Statistician* 34, 81–93.

Wainer, H., and Schacht, S. (1978). "Gapping." *Psychometrika* 43, 203–212.

Wainer, H., and Thissen, D. (1981). "Graphical data analysis." *Annual Review of Psychology* 32, 191–241.

Wald, A. (1980). "A method of estimating plane vulnerability based on damage of survivors," CRC 432, July 1980. (These are reprints of work done by Wald while a member of Columbia's Statistics Research Group during the period 1942–1945. Copies can be obtained from the Document Center, Center for Naval Analyses, 2000 N. Beauregard St., Alexandria, VA 22311.)

Walker, F.A. (1894). *Statistical Atlas of the United States Based on the Results of the Ninth Census.* Washington, DC: Bureau of the Census.

Zabell, S. (1976). "Arbuthnot, Heberden and the *Bills of Mortality.*" Technical Report #40. Department of Statistics, The University of Chicago.

Credits

Grateful acknowledgment to reprint the indicated figures is given below to the following people and organizations.

Section I
The New Yorker magazine for figure 5.

Chapter 1
Edward R. Tufte for figures 8, 9 and 10; The American Statistical Association for figures 3, 30, 31, and 32; *The New Yorker* magazine for figure 1.

Chapter 2
William S. Cleveland for figures 1, 2, and 3.

Chapter 3
Edward R. Tufte for figure 4; The American Association for the Advancement of Science for figure 2.

Chapter 4
Joe Aldenfer for figure 2; The University of Chicago Press for figures 2 and 3.

Chapter 6
Edward R. Tufte for figures 1 and 2.

Chapter 8
The *Ottawa Citizen* for figure 8.

Chapter 9
Forbes magazine for figure 3.

Chapter 12
John Wiley & Son Inc., for figure 1; The Royal Statistical Society for figures 5 and 6.

Chapter 13
Computers in Physics for figure 2; Jacques Bertin and the University of Wisconsin for figure 1; Robert Sherman for figure 3.

Chapter 15
Hyperion Press for figures 1, 2, 4, and 5.

Chapter 18
William S. Cleveland for figure 1.

Index